Man Versus Microbe

What Will It Take to Win?

Man Versus Microbe

What Will It Take to Win?

Brian Bremner
Bloomberg L.P.

NEW JERSEY • LONDON • SINGAPORE • BEIJING • SHANGHAI • HONG KONG • TAIPEI • CHENNAI • TOKYO

Published by

World Scientific Publishing Europe Ltd.

57 Shelton Street, Covent Garden, London WC2H 9HE

Head office: 5 Toh Tuck Link, Singapore 596224

USA office: 27 Warren Street, Suite 401-402, Hackensack, NJ 07601

Library of Congress Cataloging-in-Publication Data
Names: Bremner, Brian, author.
Title: Man versus microbe : what will it take to win? / Brian Bremner.
Description: Hackensack, NJ : World Scientific, [2022] |
Includes bibliographical references and index.
Identifiers: LCCN 2021062277 | ISBN 9781800611139 (hardback) |
ISBN 9781800611207 (paperback) | ISBN 9781800611146 (ebook for institutions) |
ISBN 9781800611153 (ebook for individuals)
Subjects: MESH: Microbiological Phenomena | Biodiversity | Biosecurity
Classification: LCC QR41.2 | NLM QW 40 | DDC 579--dc23/eng/20220118
LC record available at https://lccn.loc.gov/2021062277

British Library Cataloguing-in-Publication Data
A catalogue record for this book is available from the British Library.

For any available supplementary material, please visit
https://www.worldscientific.com/worldscibooks/10.1142/Q0329#t=suppl

Typeset by Stallion Press
Email: enquiries@stallionpress.com

Printed in Singapore

To Yuki, Marie & Elena, the three complex, eukaryotic life forms who bring so much joy to my personal biosphere.

About the Author

Photographer:
Natalya Chagrin

An award-winning journalist and author, **Brian Bremner** is the Senior Executive Editor for Global Business at Bloomberg News. For the past two years, he has overseen the news organization's worldwide coverage of the COVID-19 pandemic, has written extensively about the crisis and co-produced a documentary on the failed global response to the outbreak.

Bremner won an Overseas Press Club of America Award for his coverage of the Asian Financial Crisis, as well as other accolades for his reporting on Japan, India and China. He has held senior editorial posts with Bloomberg, and prior to that at *Bloomberg Businessweek* magazine, in Tokyo, Hong Kong, New York and London, where he is currently based.

Contents

Introduction

On March 20, 1995, a younger version of myself descended into the Tokyo subway and headed toward an office tower in Kasumigaseki, the Japanese capital's government quarter. I had just started as a correspondent at *Business Week* magazine, covering the nation's stressed financial system after the collapse of massive stock market and real estate speculative bubbles. One station stop before my final destination, we were told to evacuate the train.

That morning a doomsday cult known as Aum Shinrikyo (Aum) had staged a synchronized attack on the world's busiest metro subway system that carries seven million passengers a day. Aum members had planted and punctured plastic bags full of a chemical warfare nerve agent called sarin on multiple train lines all converging on Kasumigaseki Station. With its ministries, nearby Japanese National Diet Building, and Prime Minister's Official Residence, the bustling area is the Japanese equivalent of Capitol Hill in Washington, D.C. or Westminster in London. It was a target-rich environment for terrorists.

Back on street level and edging closer to my office, I encountered a chaotic scene as paramedic crews frantically treated seriously ill morning commuters. First responders, some in hazmat suits, ventured down into the subway lines in central Tokyo to rescue other passengers. In all, thirteen people ultimately died, with more than five thousand sickened, in a macabre act of violence that stunned counter-terrorism experts worldwide.

It later transpired that the chemical attack might have been a backup plan. Aum had first spent years, and considerable financial resources, trying to aerosolize and disperse a biological weapon, *Bacillus anthracis,* the bacterium that causes anthrax, a highly lethal disease.

Aum Shinrikyo, which translates into Supreme Truth, was founded and led by a semi-blind acupuncture and Chinese medicine practitioner named Shoko Asahara, who melded tenets of Hinduism and Buddhism with a nihilistic vision of a coming apocalypse. The movement attracted well-educated followers in Japan and abroad, particularly in Russia.

Several of the cult's top leaders had science and engineering backgrounds, and the head of its bioweapons program studied gene therapy at Kyoto University's viral research center. Aum had also managed to procure VX nerve gas and a Russian military helicopter, and had dispatched followers to Africa in an attempt to secure samples of the virus that causes Ebola, a hemorrhagic and deadly disease.

Despite years of effort, Aum never managed to weaponize its biological agents. Even so, bioterrorism experts today believe Aum pulled together one of the most ambitious biological weapons programs ever by a non-state actor.[1]

Micropian Age

Two months after Aum's brazen strike, the geneticist and biochemist J. Craig Venter and his colleagues published a paper announcing the first completely sequenced genome of a self-replicating, free-living organism, the bacterium *Haemophilus influenzae* Rd., which causes respiratory infections and meningitis in children.[2]

Venter and his team were looking for a proof-of-concept affirmation of "shotgun" genetic sequencing, then a new technology that used powerful computing programs to sort random pieces of genetic information into the complete genome of an organism. At that point, only a few simple viral genomes had been mapped out. So decoding this more complex microbe, short-handed as *H. Flu*, was significant. It ushered in a revolutionary shift in the study of microorganisms, or microbes.

By turns a brilliant innovator, dazzling showman and outspoken provocateur, Venter was a prime mover behind a genetics awakening in the 1990s. A self-described surfer wannabe and Vietnam vet, who worked as a medic at a Danang hospital, an experience he described as the "University of

Death" and which triggered a suicide attempt, Venter's combativeness and risk-taking won him a reputation as a swashbuckling, scientific visionary.[3]

His work on *H. Flu* was the leading edge of a string of discoveries that helped scientists begin to unravel the genetic secrets of microbial DNA, short for deoxyribonucleic acid. With its two twisting paired strands, this chemical compound contains instructions needed for self-replicating, cellular organisms to function. It's the source code of most complex biological life.

Venter would go on to play a crucial role in sequencing the entire human genetic code, or genome. After a clash over strategy with the international government consortium running the Human Genome Project, he founded a company called Celera Genomics to sequence the genome in the private sector. The resulting competitive race accelerated research in the field, and by 2003 roughly 20,000 protein-coding genes that run our bodies had been sequenced, a landmark scientific achievement that would later transform medical and drug research.

In the early 2000s, Venter circled the globe in his luxury yacht *Sorcerer II*, as part of a global scientific mission to identify new viruses and bacteria and eventually genetically detail Mother Earth's genome. The voyage drew comparisons to Charles Darwin's fabled travels on the H.M.S. *Beagle* to the Galapagos Islands and Great Barrier Reef that informed his theory of evolution unveiled in the book *On the Origin of the Species.*

Then came one of Venter's most memorable legacy plays. In 2010, his team at the J. Craig Venter Institute, a research entity he created, built the first synthetic life form. Venter's research team did so by assembling the genome of a goat pathogen called *Mycoplasma mycoides* from pieces of DNA built in a lab.[4]

Scientists then implanted the bioengineered genome into another bacterium. The transplant worked and fired up a newly created hybrid cell, which in turn started replicating over and over again in an act of molecular genesis. Venter has described the emerging field of synthetic biology as "a very important philosophical step in the history of our species. We are going from reading our genetic code to the ability to write it."[5]

That frisson of unbounded optimism, the abiding sense that humanity was on the cusp of an amazing run of biological discoveries that would extend our lifespans and enrich our societies, would be reaffirmed anew in 2012.

In one of the biggest discoveries in the history of modern science, teams led by Jennifer Doudna and Emmanuelle Charpentier outlined the biochemical mechanisms behind a new gene-editing technology called CRISPR-Cas9, which is modeled in part on how bacteria copy and cut the DNA of predatory viruses called bacteriophages seeking to annihilate them.[6]

Years of research by a variety of scientists into the chemical warfare between microbial bacteria and viruses had yielded insights that created a powerful new tool for humanity that could add, alter, or replace sections of a DNA sequence. Thanks to its pinpoint accuracy and low costs, CRISPR could theoretically be employed to boost agricultural productivity, repair genetic disorders like sickle cell anemia and hemophilia, and even contribute to breakthroughs in the quest to cure humankind's biggest killers: cancer and heart disease.

Doudna, a biochemistry professor at the University of California, Berkeley, and Charpentier, a microbiologist and geneticist who founded the Max Planck Unit for the Science of Pathogens in Berlin, received the 2020 Nobel Prize in chemistry in recognition of the transformative impact of their gene-editing tool that makes it "possible to change the code of life over the course of a few weeks."[7]

Small Things Considered

This book is about humanity's competitive, symbiotic and precarious relationship with the microbial world, why we urgently need to better understand how it works and the existential risks that we face if we fail. Our understanding of microbes — their amazing biodiversity and how they shape biological life and Earth's terrestrial, oceanic and atmospheric systems — has changed profoundly over the past two decades.

In the century ahead, unraveling the secrets of this infinitesimal dominion could dramatically improve human health and food productivity, repair our damaged oceans, and perhaps even lessen the destructive impact of climate change. Scientists, technologists, entrepreneurs, and biosecurity experts worldwide are searching for ways to better coexist with this pulsating, unseen parallel universe all around us.

Darker outcomes, less uplifting scenarios, may also be part of this scientific journey. There are viral and bacterial threats several mutations away from jumping into human populations and wreaking chaos. Our carelessness in managing dual-use technologies and open-source access to sequenced genetic blueprints of dangerous microbes and toxins is a potential opening to aspiring bioterrorists, dreaming up ways to inflict societal harm in pursuit of political or nihilistic aims. The world's current biosecurity framework has considerable vulnerabilities in an age when the ability to redesign cells and print DNA is becoming ever more accessible. We'll need to be wary of next-generation Shoko Asaharas.

This renaissance in microbiology has allowed us to learn more about the extraordinary superpowers of these life forms, which survive in extreme environments — from scalding hydrothermal vents in deep ocean trenches to the International Space Station — that would crush most biological entities.

They eat rocks, breathe metals, and break down oil spills. The first of them, bacteria, arrived well before us, perhaps as far back as 3.5 billion to 4 billion years ago. Modern humans are approximately 200,000 years old. The genus of the *Homo* species, from which we *Homo sapiens* emerged, has been around all of 2.5 million to 3 million years.

At the same time, new fields such as metagenomics and bioinformatics have given scientists a more nuanced view of the broader microbial universe. Advanced sequencing techniques, combined with high-powered computational analysis, have opened the way to analyze the key information-carrying molecules in DNA in a teeming and diverse population of microbes.

Instead of looking at one class of microbes cultured in a lab in isolation, metagenomics offered a far more expansive and dynamic look at how these tiny creatures collaborate cross-species in the towering, shape-shifting ecosystems that power core feedback loops in our land, oceans, atmosphere and human bodies.

If there's life elsewhere in the universe, many scientists think it will most likely be microbial, rather than some super-advanced, killing machine favored by Hollywood scriptwriters. Astrobiologists are looking for evidence that microbes once existed in the primordial waters on Mars, or might now be residing in a subterranean ocean on Europa, one of Jupiter's moons. Back here on Earth, it's clear that we need microbes way more than they need us. If their world collapses, so does ours.

Supersystem of Life

Given their primacy in Earth's biosphere, decoding microbial secrets will be one of the most urgent scientific pursuits in the century ahead. Even with the remarkable discoveries in recent decades, there remain considerable blind spots in our knowledge, potentially dangerous ones.

The scale and complexity of microorganisms defy easy comprehension, even for the experts. In the taxonomy of living things, biologists categorize organisms into super kingdoms or "domains." Microbes fully take up two of them, Bacteria and Archaea, and they are well represented in the third called Eukarya.

Bacteria are among the first single-cell life forms to emerge on Earth and dominate the biosphere. One rough estimate is that there are five million, trillion, trillion of them.[8] Scientists are in general agreement that life descended from a Microbial Eve, an ancient single-cell, bacterium-like organism called the Last Universal Common Ancestor, that existed roughly four billion years ago. Back then, the Earth was all of 500 million or so years old.

In 2016, a team of scientists updated the "tree of life" (Fig. 1), using the latest computational techniques. They analyzed genomic data to show not only the evolution of biological life, but also the distribution of its biodiversity. The life forms that we can see such as plants, animals and humans represent a very tiny slice of the evolutionary map. In this rendering of life, Bacteria make up two-thirds of all biological diversity and Archaea nearly a third.[9] We're living in a microbial world.

Bacteria come in spherical, rod-shaped and spiral varieties. Many work cooperatively with plant and animal life; others cause some of the world's most lethal infections such as anthrax, tuberculosis and syphilis. They can survive in environments from human digestive tracts to superheated crevices deep in the Earth's crust and radioactive waste.

Archaea are also varied and adaptable ancient microbes, but with a uniquely different molecular structure and didn't become an officially

Figure 1: In this tree of life, various types of Bacteria (upper left, center and right) and Archaea dominate. Eukaryotes, including humans, are a strand of the top stem on the lower right. Image credit: Graphic by Zosia Rostomian, Lawrence Berkeley National Laboratory.

recognized domain until the late 1970s. Like many microbes, they tend to coexist transactionally with other life forms, providing a beneficial service in exchange for something of value to them. They are incredibly agile, too, able to use inorganic chemical compounds like carbon dioxide and ammonia to generate energy and organic material, and play a key role in the global food web.

Then there is the Eukarya, the third domain of more complex and often visible living things like plants, animals, and Lebron James. It also contains the microbial kingdoms, a subcategory of a domain, of Fungi and Protista. You're probably familiar with fungi, which includes yeasts, molds, mildews and mushrooms.

Protists are more versatile. They include giant kelp, the world's largest seaweed and marine algae, that grow as long as 30 meters (about 98 feet) and form underwater forests that provide homes for scores of marine species.

The protista world is also home to microalgae life forms such as diatoms and dinoflagellates capable of photosynthesis, or converting sunlight into chemical energy. Some dinoflagellates are bioluminescent, meaning they actually generate light.

Viral Zombies

Microbes within the domains of Bacteria and Archaea are called prokaryotes (pronounced like pro-ka-ree-oats), single-cell organisms with few internal structures and no nucleus. The Eukarya consists of eukaryotes and are generally more complex. Their cellular structure typically does have a nucleus, where their DNA information is stored, and the ability to bolt together to form complex, multicellular life.

Viruses are bizarre outliers, and they're not officially counted in the three biological domains. They're the ghoulish undead of the vast microbial menagerie, opportunistic pieces of genetic material that come alive only when they hijack the cellular machinery of a host. They are among the least understood biological entities on Earth.

Viruses are the smallest of the microbial small (500 million common cold rhinoviruses fit on the head of a pin) yet what they lack in size, they make up for in numbers. If you stacked all the estimated viruses end-to-end, they would stretch out 100 million light years.[10]

In the pantheon of microbial threats, viruses have an especially villainous image, given their role in dreaded maladies like smallpox, rabies, Ebola and AIDS (acquired immunodeficiency syndrome). In our oceans, viruses are locked in mortal combat with bacteria, devouring them in a non-stop microbial conflagration.

And yet, most are quite harmless to humans and their rapid-fire rates of mutation make them nature's innovation labs, creating entirely new genes crucial for evolutionary dynamism. Ancient viral genetic information is embedded in our human DNA and may play a role in warding off some diseases. One type of virus known as a bacteriophage specializes in devouring bacterium, some of which cause health-threatening infections.

As for the overall size of the microbial world, there are an estimated one trillion known individual species of microorganisms on Earth, based on very rough extrapolations of DNA samples. Yet microbiologists believe that many multiples of that number have yet to be discovered. We honestly don't have a clue about the exact number of microbes actually out there.

When microbes commune in great numbers (they represent 90 percent of the biomass of our oceans), these microscopic life forms exhibit an amazing collective intelligence, honed over billions of years of evolution. They communicate, interact collaboratively and solve problems. We just don't know yet how human industrial and agricultural activity, the dramatic loss of animal and plant biodiversity, or climate change are altering the microbiome on a macroscale.

What happens, for instance, if microbial life goes into overdrive in the Arctic region's thawing permafrost zone. It's one of the world's biggest carbon storage systems, thanks to plant and animal organic material scraped up by glaciers receding after the last ice age 11,000 years ago and frozen for millennia.

There's now about twice as much carbon stored in permafrost than is currently in the atmosphere in the form of greenhouse gases. As the planet warms and that frozen ground thaws, some climatologists and microbiologists fear runaway carbon emissions, as microbes move in to feast on newly available organic material, dramatically expanding the amount of climate-warming, methane gas emissions in the Earth's atmosphere.

Amazingly, the supercomputer-powered climate models that guide world leaders and the public on the trajectory of climate change don't fully account for the role that microbes play in recycling greenhouse gases such as carbon dioxide, methane and nitrous oxide. Until that changes, the risk of an abrupt and chaotic climate breakdown later in the century may be far higher than commonly acknowledged by world leaders.

At the same time, fungi appear to be busting out of climate zones they've long inhabited. Pathogenic varieties of *Candida auris* have rapidly spread worldwide in recent years, and are starting to be a serious public health threat, preying on hospital patients with weakened immune systems.

In 2019, the world's leading microbiologists issued a rallying cry for humanity to get smarter with regard to the intricate role that microbes play in climate change. In a consensus statement, these notables placed "humanity on notice that the impact of climate change will depend heavily on responses of microorganisms, which are essential for achieving an environmentally sustainable future."[11]

Species Malpractice

If the world needed a reminder of how vulnerable we are to microbial threats, a novel coronavirus called SARS-CoV-2 that first appeared in China delivered one with the subtlety of a chainsaw. The global health emergency of the early 2020s has been called a twenty-first-century "Chernobyl moment" and the "worst combined health and socio-economic crisis in living memory, and a catastrophe at every level" in a report by an independent panel appointed by the World Health Organization.[12]

Our current strategy of panicking when infectious disease crises surface, then returning to a posture of studied inaction once they've subsided, could consign humanity to wave after wave of destabilizing outbreaks for decades to come. COVID-19 has killed roughly six million worldwide as of early 2022, rocked the global economy and disrupted lives worldwide.

With epidemiological ripple effects from the pandemic likely to linger for years, economists see the ultimate financial hit reaching $22 trillion in the 2020–2025 timeframe, according to the WHO study. That's the biggest shock to the global economy since World War II and the most severe simultaneous contraction of national economies going back to the Great Depression of the early 1930s. Consequently, anywhere from an estimated 115 million to 125 million people have been pushed into extreme poverty.

The ultimate origin of that virus is still unknown more than two years after its first outbreak. Theories range from a species jump from the bat caves of Central China or Southeast Asia to an accidental leak at the Wuhan Institute of Virology, home to a biosafety level 4 lab, whose work with infectious pathogens requires the highest level of security. We don't know the origin of the virus in part because the Chinese Government has limited access to international investigators.

Regardless of the source, this was far from a random, Black Swan event. In late 2002, another virus from the same coronavirus family stormed out of

China's southern Guangdong province and killed about 774 people from Asia to Canada with an airborne illness, dubbed severe acute respiratory syndrome (SARS).

That crisis proved short-lived. Ironically, eighteenth-century techniques like contact tracing and quarantines probably played a bigger role containing the infection than modern science. The virus burned itself out without the development of a vaccine. However, virologists viewed the episode as a wake-up call and warned policymakers for years that the SARS outbreak was merely a dress rehearsal for bigger contagious calamities to come. Then COVID-19 happened.

As disorienting and destabilizing as the COVID-19 crisis has been, it could have been a lot worse if SARS-CoV-2 were more lethal or molecularly complex. Catastrophic risk modelers like the San Francisco-based infectious disease data and consulting firm Metabiota that work with insurers and governments have mapped out scenarios for various microbial infections depending on their transmissibility, case-to-fatality ratio and time needed to develop an effective vaccine. The average number of expected deaths were 264 million globally, ranging from 117 million to as high as 545 million, according to the 20 worst-case, multi-year pandemic scenarios simulated by Metabiota's models.

Looking back, the unbridled optimism about the microbial world of recent decades feels very different today in the midst of a still-simmering global health emergency. If you consider the most complex and looming civilizational challenges facing humanity in the next hundred years such as pandemic risk, food security, ocean health, bioterrorism and climate change, the microbial domain is intrinsically linked to all of them.

Over the past two decades, we've seen an increase in infectious disease outbreaks that jump from livestock and wild animal hosts (from birds and swine to bats and mosquitoes) to humans. Since the 1970s, more than 1,500 new potentially disease-causing pathogens, 70 percent of which were of animal origin, have been discovered.[13] Virologists and epidemiologists concede that we are a long way away from full situational awareness of all the microbial disease threats lurking in nature.

The disease spillover risks are increasing, as population pressures and agricultural production destroy natural habitats like rainforests, grasslands and other wildlife reservoirs. Satellite images reveal that nearly 40 percent of the world's land surface is now devoted to crop production and livestock, as we strive to deliver protein and calories to 7.9 billion people. That number will be closer to 10 billion in 2050 and 11 billion by 2100.[14]

All this not only creates more avenues for animal viruses and bacteria to leapfrog species, but also may be altering microbial ecosystems that recycle carbon, nitrogen and nutrients that keep our soil productive. Overuse of antibiotics in livestock production has created a new breed of drug-resistant superbugs that by 2050 are expected to kill more people annually than cancer.

Microbes are one of Big Science's unknowns. We're rather like Neo early on in the 1999 science fiction classic *The Matrix*, stumbling around in a world that's illusory at some imperceptibly deeper level. Neo ultimately discovers that robotic sentinels have enslaved humanity. The sentinels shaping our destiny aren't malevolent and utterly indifferent to our fate. They just want to *be*. We need to figure out how to coexist.

Biohacking Armageddon

The real microbial threat might ultimately be us. It's true that naturally occurring pandemics have certainly taken their toll on humanity. Smallpox killed an estimated 300 million in the twentieth century.[15] That's more than three times the combined global fatalities from World War I (31 million) and World War II (58 million).

Sometimes, we catch a break from Nature. Infectious diseases like Ebola or Nipah have high case-fatality levels, but relatively low reproduction rates. Super-transmissible maladies like measles and whooping cough have fairly low fatality rates. From an evolutionary perspective, disease-spawning viruses or bacteria have more staying power if they don't kill their hosts too quickly.

A well-designed bioweapon (or a lab security breach of a super-strain) would be a force multiplier. In 2012, a team of researchers led by Yoshihiro Kawaoka, microbiologists with the University of Wisconsin–Madison, published a paper in *Nature* outlining how they genetically modified the H5N1 avian influenza into contagious forms that could spread among ferrets, a mammal proxy for humans, via respiratory droplets.[16]

Advocates argue that these so-called "gain of function" experiments are a crucially important part of infectious disease research, allowing scientists to pinpoint causal relationships between genes or mutations and to better understand the characteristics of a microbial pathogen. Others fear an accidental lab security breakdown that unleashes something ghastly into the environment.

In 2018, two Canada-based virologists published a study showing how they synthesized the horsepox virus, a relative of smallpox, from genetic fragments for roughly $100,000 using a mail-order DNA supplier. In defending their work and decision to publish, Ryan S. Noyce and David H. Evans argued the potential benefits of this kind of research outweigh the risks: "Synthetic biology offers enormous promise as a tool for engineering advanced biotherapeutics" for such diseases as malaria, human immunodeficiency virus (HIV) and hepatitis C virus (HCV), they wrote. "Realistically, all attempts to oppose technological advances have failed over centuries."[17]

True enough, yet in the early 2020s more traditional biotech research is being outsourced to automated "cloud labs," while artificial intelligence and machine learning applications do deep analysis of thousands of molecules. That's expanding the opportunities for bad actors to do serious damage. With a little bit of effort, you can find the fully sequenced genome of the smallpox virus on the Internet.

Counter-bioterrorism leaders must now worry about more potential actors than just state-sanctioned programs by geopolitical adversaries. A biohacker could steal sensitive biotech and sell it on the dark web to the highest bidder. A disenchanted researcher with access to corporate or university DNA synthesizers that "print" strings of genetic information could also emerge as a future bio-anarchist.

Alpha and Omega

Gaze out into the middle of this century and beyond, and it's clear humanity has much more to learn from microbes. Will synthetic biology allow us to extend our lifespans or lead us into a nightmarish dystopian era? Could carbon-eating bugs be programmed to save the planet from overheating? Or will a carbon bomb emission explosion detonate on Earth's thawing rooftop? How do we protect ourselves from future pandemics and bioweapons?

Microbes are the alpha and omega of our most profound scientific and philosophical questions about the origin of life and the future of the species. Our creation story is microbial, though we still don't know the exact biochemical chain of events that created the first ancestral microbe and spawned more complex life forms.

Are we alone in the universe? Or will we find evidence of microbial life elsewhere? And what happens to life on Earth when billions of years from now, a dying sun is expected to expand into a red giant and vaporize all life on the planet?

That's an unimaginable time span. Yet might these tiny life forms, so well-suited to long-haul space travel, be our ultimate star-death insurance policy? Could microbes be our future interstellar ambassadors to seed life elsewhere or be humanity's witness statement that we existed at all?

There's much to think about as we get our minds around the unseen miniaturized reality that surrounds us. And we will explore the risks, opportunities and deep future scenarios that humanity's uneasy relationship with microbes poses in the three major sections of this book.

First we need to consider this: Why are we so bad at managing long-term, probabilistic risks like pandemics? We know emerging infectious pathogens, bioterrorism vulnerabilities and other microbial threats loom. Something is holding us back. It might be our Stone Age brain circuitry.

THE RISKS

Chapter 1

Global System Failure

When it comes to navigating the government-science nexus or calibrating the political agenda of the White House, few are as savvy as Dr. Anthony Fauci, America's top infectious diseases authority and director of the National Institute of Allergy and Infectious Diseases (NIAID).

Fauci has served under seven presidents, winding back to Ronald Reagan. He has had a backstage pass to some of the biggest infectious disease dramas of the last four decades, overseeing research into the human immunodeficiency virus (HIV), tuberculosis, malaria, Ebola, and Zika.

The son of Italian immigrants, Fauci grew up in an apartment above his family's pharmacy in Brooklyn, New York. From there, he made his way to what was then called the Cornell University Medical College, graduating first in his class. Over the years, the immunologist has emerged as one of his field's most cited researchers for his work into how HIV disables the body's natural defenses and triggers a pernicious disease called acquired immunodeficiency syndrome, or AIDS. He's also done important research on other immune system diseases.

If there's such a thing as a global scientific celebrity, Fauci comes close. In 2008, President George W. Bush bestowed upon him one of America's highest civilian honors: the Presidential Medal of Freedom. In doing so, Bush praised Fauci for a public service career that expanded "the horizons of human knowledge."[1]

In 2020, and in a very different America, Fauci found himself on the receiving end of death threats, credible enough to require a Secret Service security detail. His adult children were menaced, too, at home and at work. Someone sent Fauci a hoax letter that when opened covered him in a puff of white powder, presumably meant to simulate a biological toxin like anthrax or ricin.[2]

As wave after wave of COVID-19 infections killed hundreds of thousands of Americans and undermined his hold on power, President Donald Trump called Fauci a "disaster" and an "idiot" in a newspaper interview.[3] One former Trump adviser suggested that Fauci should be beheaded;[4] another called him a "liar and a sociopath,"[5] and falsely suggested that one of America's most decorated public health authorities had helped China genetically engineer the virus.

America's faltering response to the SARS-CoV-2 outbreak was startling, even rather humiliating. A country that produced a world-leading number of Nobel laureates, led the development of the first polio vaccine and, yes, put a man on the moon, had become a coronavirus superpower, leading the world in infections and fatalities. With nearly one million deaths as of the end of February, 2022, the U.S. had one of the highest COVID-19 deaths per capita in the developed world. The country's deadliest pandemic has resulted in more loss of human life than the combined American deaths during the last century's two world wars.

In a rapidly evolving crisis, America's scientific establishment stumbled. The U.S. Centers for Disease Control and Prevention (CDC) released a faulty diagnostic test in early 2020 that set back efforts to track the virus. Stockpiles wound down quickly, followed by shortages of facemasks, gloves, gowns, and ventilators. Fauci and U.S. Surgeon General Jerome Adams initially discouraged the public from buying masks in late February and early March of 2020 due to concerns about shortages for healthcare workers.

A month later, as more data showed the virus could spread stealthily in individuals showing no symptoms, the CDC and health officials changed their stance and recommended face coverings. Trump only half-heartedly backed the idea, rarely wore masks in public and repeatedly downplayed the risks posed by COVID-19 or prematurely raised hopes the crisis would subside quickly. The resulting public confusion about COVID-19 guidelines never entirely lifted.

Fauci had certainly faced public backlashes before. In 1984, as the newly-promoted director of the NIAID, which is part of the National Institutes of Health (NIH), he endured withering criticism over the Reagan

Administration's perceived indifference to the emerging AIDS calamity that ravaged gay communities in San Francisco and New York.

President Reagan famously didn't acknowledge the epidemic until 1985, roughly four years after the viral contagion, some derisively called the "gay plague," took off globally. As the leading and public-facing scientist in the AIDS response, Fauci took more than his fair share of slings and arrows. Protestors burned him in effigy during protests outside NIH headquarters in a suburb of Washington, D.C.

From a purely epidemiological perspective, HIV is a formidable microbial adversary. Its molecular clock is blisteringly fast, churning out mutated copies of itself that recombine into new configurations as the virus moves from one host to another. According to one study, it exhibits "the highest recorded biological mutation rate currently known to science."[6]

The virus also targets and disables specialized white blood cells called CD4 T lymphocytes that play a crucial role in the immune system's ability to identify and kill invading pathogens. Most perniciously, once HIV is circulating in the bloodstream it creates a subset of the virus called a provirus. These sleeper proviruses are integrated into the DNA of healthy cells and can be activated later. HIV's ability to conceal itself from our immune system, or the drugs meant to neutralize it, has confounded AIDS vaccine researchers for nearly four decades.

Specialists are still pursuing an effective vaccine for AIDS, and a combination of antiretroviral medicines now allows many of those infected with the virus to live normal lifespans. That said, this viral microbe has killed an estimated 36.3 million worldwide.[7]

As traumatic as HIV has been in the annals of global infectious disease outbreaks, Fauci told me that he thinks COVID-19 will ultimately leave a far bigger imprint on the world. Untreated AIDs is basically a death sentence, but HIV was never as highly transmissible as an airborne respiratory virus like SARS-CoV-2 that can strike you any time, anywhere.

AIDS wasn't a "diffuse outbreak in the sense of generally being restricted to certain demographic groups, in certain countries," Fauci said. In contrast, COVID-19 was a "respiratory illness that threatens every person on the planet, shuts down the economy of virtually every country. We are living through really a historic experience."[8]

The COVID-19 cataclysm occurred in a world far different than that of the early 1980s. The open societies of the West are now more politically polarized and wealth inequality has deepened over the decades.

This was also the first pandemic of the digital age, with social media networks that served as superspreaders of misinformation about the virus and vaccines that reached billions of users worldwide. The broader media landscape was fragmented and politically balkanized.

The resulting information bubbles meant that there was no single reliable narrative about what risks the pandemic posed or what type of precautions one should reasonably take. Some Americans, even as the viral outbreak killed their fellow citizens in their local community, perceived the virus as a "deep state" conspiracy by elites like Fauci and corporate media, pursuing their own political and financial agendas.

"The depth of the hostility to science truly surprised me," Fauci mused. "There were certain regions of the country, where in the very towns, cities, and counties, where the hospitals were being overrun with COVID-19 patients, there were people saying it's a hoax, it's fake news, it doesn't exist. We're going through a truly historic pandemic that people feel is not real."

Stone Age Minds

The level of pandemic denial in regions of the United States, as well as other parts of the world spanning from Eastern Europe to Latin America, seems surreal. Yet it actually isn't all that surprising, according to evolutionary biologists and cognitive psychologists, whose research suggests that humanity has evolved a very specific set of behaviors that shape our response to contagions.[9]

Our early ancestors lived in nomadic hunting groups and faced the omnipresent threats of starvation, predation, parasites, and violent raids by rivals. Nature, as mid-nineteenth-century British poet Alfred Tennyson put it, was "red in tooth and claw." Contemplating a statistically abstract and worldwide event like a pandemic was not a pressing concern 200,000 years ago, as the brains and emotional systems of modern humans were starting to evolve.

If you look at the history of infectious disease outbreaks, across millennia and technological eras, the way societies have reacted to them is remarkably similar. We're more interested in individually avoiding pathogens than developing rational group strategies to prevent and contain them; we're more likely to scapegoat outsiders as a crisis intensifies; we're more susceptible to conspiracy theories that confirm our beliefs, justify our actions or deny harsh realities; and we're more tempted to "free ride" off the sacrifices of others, if we calculate that we will go unpunished.

Scientists with an evolutionary perspective claim that we've adapted disease-avoidance behaviors that, while making sense at an individual level, can create widening gyres of social disorder for the community at large. These ancient mechanisms, rooted in our early history as a species, remain embedded in our neural networks and emotional systems to a surprising degree.

Unfortunately, our Stone Age brain circuitry doesn't match up well against a geometrically expanding viral or bacterial pathogen in an era of international travel, planet-spanning supply chains and split-second global communication. In a rapidly evolving crisis, humans aren't always rational. That gives opportunistic microbes in search of warm-body, nutrient-rich hosts a huge advantage.

The theory that all biological organisms undergo cumulative changes over time, thanks to random mutations and natural selection is, of course, primarily credited to Charles Darwin. It's one of the most powerful insights in the history of science, though, as we shall explore later, some of traditional evolutionary theory's most cherished assumptions are now being challenged by new insights into how microorganisms swap genes in vast colonies called microbiomes.

Darwin viewed evolution essentially as a struggle for existence in a world of finite resources. Over lengthy time spans, subtle, random changes in our biology take place. Species that undergo changes better adapted to one's environment are more likely to survive, reproduce and pass these characteristics to offspring. Less adaptive species fade into oblivion.

At the molecular level, a virus or bacteria that can evade a human immune system, bind to and hijack healthy cells, replicate efficiently and jump from species to species will thrive. Humans have evolved defenses, too, both biological and behavioral. We have sophisticated pathogen-targeting cellular networks as part of our immune system.

Our perceptual systems, the way we take in and process information about our environment, are a product of evolutionary natural selection as well. We've evolved strategies and capabilities to find food, avoid predators and parasites, as well as to attract mates.

The biological immune system destroys most microbes that may do us harm, unless of course it's a novel pathogen like SARS-CoV-2 that it hasn't seen before. Yet in metabolic terms, immune responses are expensive, requiring a lot of energy or calories to turn on and run. Fighting diseases wears us down, makes us feverish and fatigued.

So our ancient ancestors may have also evolved what evolutionary psychologists call the *behavioral immune system*, a more proactive set of adaptations akin to preventative medicine.[10] We gag at the sight of rotting food or a decomposing animal on the side of the road. Numerous studies have shown that disgust is an adaptive system for disease avoidance behavior.

All manner of species have evolved similar strategies. Mammals and birds remove parasites from their furs and plumage via grooming and preening. Experiments reveal that lab mice avoid fellow comrades whose immune systems have been activated after receiving an injection of a pathogenic extract. Even bacterial microbes have developed an adaptive immune system to ward off predatory viruses.

When triggered in humans, our behavioral immune systems have big implications for the way we act in groups, our level of social gregariousness, our view of outsiders, and our willingness to conform to community sacrifices and expectations in an emergency.

As humans started to migrate from nomadic hunter-gatherer groups to larger, permanent settlements that tended to crops and livestock approximately 10,000 years ago, the environment for opportunistic pathogenic microbes turned far more hospitable. The comingling of humans and domesticated animals offered pathways for predatory viruses and bacteria to find new hosts. We would learn that painful lesson over and over again.

From Athens to Wuhan

Had Thucydides, one of the ancient world's greatest historians, been around to witness the COVID-19 pandemic that destabilized the world in the early

2020s, he might have been struck by the parallels with the great public health crisis of his day: the Plague of Athens (430–426 BCE).

Considered the father of political realism, the Athenian general is remembered among classical scholars for *The History of the Peloponnesian War*, which chronicled the epic clash between Sparta, the era's predominant land power, and the ascendant rival maritime city-state of Athens.

Yet scientists lay claim to this masterwork, too. It's perhaps the first detailed account of an infectious disease outbreak in recorded history, one that ravaged the Athenian population and did lasting damage to the political and cultural center of Greece. In the time of Thucydides, a "plague" was a term of art for any transmissible disease event.

Thucydides wasn't a physician by training, but did possess a keen, analytical eye and grasped the broader significance of the epidemic wave that wiped out an estimated 75,000 to 100,000 Athenian lives, or about 25 percent of the city-state's population, in nearly five years.[11]

"People were attacked without exciting cause, and without warning, in perfect health," Thucydides reported. "It began with violent sensations of heat in the head, and redness and burning in the eyes; internally, the throat and tongue were blood-red from the start, emitting an abnormal and malodorous breath. The more they came into contact with sufferers, the more liable they were to lose their own lives."[12]

The silent, invisible killer that tore through Athenian society and so terrified the Greeks remains an epidemiological mystery. Modern-day researchers speculate that smallpox, typhoid fever, or possibly the measles may have been the culprit.

It's likely that population density played a role. Facing a major land campaign from Sparta, whose warriors were feared for their skill and savagery, Athenians living in the countryside fled to the city-state's center, creating crowded conditions and an ideal setting for a predatory viral or bacterial infection to run riot.

In Thucydides' telling, the disease outbreak originated in Ethiopia, stormed around the Mediterranean region through Egypt and Libya and found its way to Piraeus, Athens' main port. From there, the pathogen spread throughout Athens, with little distinction as to social standing or class. Thucydides, himself, contracted the illness and survived.

Others succumbed to the affliction on the streets of Athens. "The bodies of the dead and dying were piled on one another and people at the point of death reeled about the streets," he wrote. "All the funeral customs which had previously been observed were thrown into confusion and the dead were buried in any way possible."

Among the most prominent victims was the great statesman and orator Pericles, under whose leadership democracy flourished, architectural wonders like the Acropolis rose, and the Golden Age of Athens transpired. With the loss of this charismatic leader, the grinding, multi-year epidemic tore at the social fabric of Athens.

"The plague marked the beginning of a decline to greater lawlessness in the city," Thucydides wrote. "People were more willing to dare to do things which they would not previously have admitted to enjoying, when they saw the sudden changes of fortune, as some who were prosperous suddenly died, and their property was immediately acquired by others who had previously been destitute."

China's Viral Crisis

Nearly 2,500 years later, a mysterious respiratory illness surfaced in Wuhan, a megacity of 11 million in central China. We had far more sophisticated technologies and knowledge of germ theory at our disposal than the ancient Greeks, yet one easily imagines Thucydides shuddering in disbelief at the quality of our global response.

After infections started to surface in December of 2019, front-line Chinese clinicians and geneticists took samples and conducted tests to identify the molecular characteristics of the novel virus. At the end of December, local officials quickly shut down a Wuhan seafood and animal wildlife market, where scores of early infections had been traced back to.

By January 11, 2020, a draft of the genome for the respiratory virus had been sequenced, uploaded and published on an open-source research database called GenBank. Efforts were quickly underway to develop PCR (short for polymerase chain reactions) tests to detect the genetic material of the pathogen.

As Chinese public health and Communist Party officialdom started to weigh in, things slowed down. In the critical early weeks of January, 2020, officials opted for secrecy and limited public health messaging, possibly to keep up appearances ahead of preparations for the annual National People's Congress, an important political ritual in which top government officials map out the nation's agenda.

Chinese doctors who sounded the alarm were muzzled, some detained. Local officials were initially reluctant to alarm the public or draw world attention to an emerging crisis. In mid-January, 2020, some 40,000 families attended an annual Lunar New Year public "potluck" banquet, an event approved by Wuhan's city officials despite an escalating number of viral infections.[13]

Days later, President Xi Jinping and the Politburo declared a national emergency. The government airlifted military doctors into Wuhan, started building a massive, temporary hospital in the city and, on January 23, cordoned off much of the central Chinese province of Hubei, virtually blockading more than 50 million citizens in the biggest large-scale quarantine in modern history.

Wuhan had gone viral just ahead of China's week-long Lunar New Year travel season that had started on January 25 in 2020, when millions of mainland tourists fanned out across Europe, Asia, and North America. It's the world's biggest annual human migration and, in retrospect, almost certainly seeded the novel pathogen overseas. While Xi locked down central China, there were no restrictions on outward international travel from other parts of China until weeks later.

Even as the World Health Organization became aware of the unfolding crisis in Wuhan starting on December 31, 2019, it was far too slow in sounding the alarm to the rest of the world. Given the history of past, fast-moving respiratory viruses, the global health guardian failed to fully assess the risk that the new pathogen spread via human-to-human contact, according to a pandemic response review by an independent panel and ordered up by the WHO.[14]

The Geneva-based health organization also decided not to declare a Public Health Emergency of International Concern during top-level meetings that began on January 22, despite the fact that the Chinese Health Ministry had just publicly confirmed person-to-person disease spread two days earlier.

By the time the WHO declared a global public health emergency on January 30, 2020, cases had been reported in nineteen countries. As of early February, the virus had reached four continents in about five weeks and was expanding exponentially. "Surveillance and alert systems at the national, regional and global levels must be redesigned," the WHO panel's review concluded.

SARS-CoV-2, as the virus is technically known, is a coronavirus, a viral family known for its crown-like, spike proteins (hence the name corona). These viral protrusions bind to a molecule, called the angiotensin-converting enzyme 2 (ACE2), that sits on the surface of many human cells. Once that happens, the virus undergoes a structural transformation that allows it to fuse into a healthy cell, commandeer its cellular machinery and start to replicate.

Coronaviruses come in multiple strains, ranging in severity from the common cold to some of the most lethal infections threatening humanity, such as severe acute respiratory syndrome (SARS) and Middle East respiratory syndrome (MERS).

Early on, there was some hope that the new virus might not be as threatening as its viral cousin SARS that had burst out of China's southern province of Guangdong in the early 2000s. Firstly, it wasn't as fatal. Secondly, while there was evidence of human-to-human spread, it wasn't clear just how efficient its transmission was in the early weeks of the crisis.

That cautious optimism crumbled as scientists started to grasp that some virus carriers had no symptoms at all, yet were still spreading the pathogen. The thermal-imaging cameras that detected travelers with fever at airports and helped contain the 2003 SARS outbreak were of less use this time around. Through much of February, 2020, the virus rampaged around the world relatively undetected and most governments were reluctant to shut down international travel.

By the time the WHO declared a pandemic on March 11, 2020, hospital systems in hotspots like Milan, Seattle, and New York were starting to face an onslaught of patients. Global shortages of masks, syringes, ventilators, diagnostic tests, oxygen supplies, and even body bags, deepened, as supply chains were overly reliant on a handful of manufacturers clustered in a few countries. Hoarding and price-gouging were the natural outcome of a global hospital equipment industry with no surge capacity for a crisis like COVID-19.

"We started to get anxious when it became clear that community spread was spread by people who had no symptoms. There may have been an extraordinary amount of infection going on under the radar screen," recalled Fauci. "Then as we got to the explosion of cases in northern Italy, in Europe, which then seeded the cases in the Northeast part of the U.S., it was absolutely clear we were in for a long, hard slog with this virus."

With fatalities rising, rich-world governments, led by the Trump Administration, allocated billions of dollars in emergency funding to accelerate vaccine development at drug and biotech companies worldwide. Toward the end of 2020, a handful of effective vaccines had been approved, thanks to crash programs by drug developers at Pfizer Inc. and BioNTech SE, Moderna Inc., and a partnership between AstraZeneca Plc. and Oxford University.

As 2022 began, much of the developing world remained unvaccinated, due to the lopsided international distribution of these drugs. Wealthier countries, representing about 20 percent of the world's population, had about twice as many doses as the rest of the world.

Despite new, and potentially more dangerous, variants of SARS-CoV-2 such as Delta and Omicron showing up in populations, experts don't see the world getting full immunization coverage until 2023 or 2024. COVID-19 also appears to do long-term damage to victims even after its infection passes. These so-called "long COVID" patients will pose healthcare challenges for millions of people in the years ahead.

Gaming the Pandemic

Why are we so bad at handling pandemics? Deadly global infectious outbreaks are low-probability yet high-risk events. They happen infrequently. Some generations are hit hard by infectious disease outbreaks; others get only a glancing blow; still others escape largely unscathed. Yet once a pathogen jumps into the broader human population pool, it tends to be with us forever.

Since the arrival of antibiotics like penicillin in the 1920s, and with improved sanitation and public health policies worldwide, humanity has made progress in reducing deaths from communicable diseases like AIDS, tuberculosis, and malaria.

However, only two diseases have been declared eradicated by the World Health Organization: smallpox caused by the variola virus and rinderpest, also known as cattle plague, from a viral microbe of the same name.[15] Granted, that's really hard to do, but neither is this a particularly impressive track record.

It's clear the world needs a faster and more comprehensive, high-tech pandemic surveillance system, more global coordination on information sharing, supply chains, drug research, and a more equitable system to distribute vaccines and treatments. In the past, we've reacted to pathogen outbreaks, rather than proactively trying to avoid them.

Again, our Stone Age brains may be slowing us down. Studies by cognitive psychologists have shown that humans are weak at long-term, probabilistic risk analysis. We aren't adept at thinking about the well-being of our future selves, let alone future generations.

Emotionally striking events in the here and now dominate our thinking far more than abstract threats in the future. Infrequent or long-term perils like pandemics, the rising risks from antibiotic-resistant bacteria or climate change's impact on the microbial supersystems running the Earth don't register with our attention scanners as readily.

Then there's what cognitive psychologists call the "availability heuristic," our predilection to imagine the future through the lens of recent events and ready information. If your perception is that climate change isn't an immediate threat now, it's harder to assess the risks in 2050 or 2100. Most of us didn't think much about pandemics until our lives were thrown into chaos in 2020.

Another cognitive bug is our predisposition to believe wrong or misleading information, which can be life-threatening in a global public health emergency. People are reluctant to entertain new information that doesn't confirm their existing beliefs (confirmation bias) or veer from a plan of action they've staked time, resources, and their reputation on (sunk cost bias).

Pandemics divide us. There's our tendency during a crisis to look for scapegoats and then justify our actions with baseless conspiracies. In the mid-fourteenth century, Jews and gypsies were tortured and burned to death for supposedly causing the Black Death bubonic plague.[16]

During the 1918 influenza pandemic, the outbreak was linked by some in the United States and U.K. to aspirin made by the German pharmaceutical company Bayer.[17] In the age of COVID-19, Asian-American hate crimes increased in both the U.K. and United States, where overall homicides spiked nearly 30 percent in 2020 and firearm sales spiraled up as well.[18]

Finally, effectively responding to a pandemic requires unified action, a citizenry willing to follow the same rules: Don't horde, don't travel, limit social interaction, wear a face mask and get a vaccine shot. That's critical in both containing infection spread and reaching "herd immunity," when a high enough percentage of the population is immunized so that the pathogen loses momentum.

Evolutionary game theory, which explores the dynamics among decision makers in a variety of settings, shows us why this is difficult in practice.[19] In contagions, the interests of the individual aren't always aligned with those of the group.

During a lockdown, everyone is asked to make lifestyle and economic sacrifices to drive down overall infection rates. If those rules aren't effectively enforced, free riders can secretly avoid that cost while still benefiting from any public health gains enjoyed by the broader community.

In so-called "public good" game simulations, over time more free riders emerge the longer the game goes on and the risk of punishment is low, eventually reaching a threshold beyond which the health profile of the entire group is undermined. Pandemics typically last years, with several infectious waves that test social cohesiveness and staying power.

In the United States, Republican-leaning states in the South and West were slow to comply with federal safety guidelines and quick to abandon them. To justify those actions, some leaders downplayed the risks or accused the public health establishment, liberal politicians or the media of self-interested scare mongering.

There's been a lot of self-serving commentary by government leaders about the relative merits of varying political models during the crisis. The Chinese state propaganda machine, diverting attention from Beijing's early missteps and obfuscations, has stressed how its policies initially restored life back to normal faster than in the West, affirming the superiority of the country's autocratic Communist Party leadership.

Leaders in the United States and U.K., which experienced relatively higher infections and death rates, have focused on how their world-class drug industries, scientific ingenuity and speedy vaccine deployments offered the world a way out of the crisis.

Before leaving office, Trump claimed at a "vaccine summit" that America's multi-billion-dollar investment in vaccine development and manufacturing would end the pandemic. British Prime Minister Boris Johnson, who spent several harrowing days in a London hospital intensive critical care unit grappling with COVID-19, suggested Britain was poised to emerge as a bioscience superpower.

There are interesting lessons to be drawn from the stunningly poor performance of Western democracies such as the U.S. and United Kingdom relative to their East Asian counterparts.[20] It turns out that government competence can save lives in a pandemic, as Singapore, South Korea, and Taiwan amply demonstrated to the world.

Cognitive psychologists tend to focus less on political models and more on a country's cultural "tightness" versus "looseness," meaning the willingness or unwillingness of its citizenry to follow rules imposed by a government and its authoritative experts in a crisis.

In countries like the United States, U.K., and much of Western Europe, there's a greater tolerance for mavericks, rule-breakers and innovators who buck established norms. In other societies, particularly in Asia, there are stricter social expectations to ensure cohesion and order. When it comes to economic dynamism and cultural creativity, looser societies might have an advantage.

In a pandemic that requires quick, coordinated action, tighter societies seem to perform better. In March, 2021, a research team that tracked more than fifty countries during the COVID-19 crisis published a study that found that, on balance, loose cultures had five times the number of cases, and eight times as many fatalities, as tight cultures did.[21]

Preemptive Strike Strategy

There are some rather obvious takeaways from global system failure during the viral outbreaks of 2020 and 2021. They include faster and more seamless digital information sharing systems, deeper stockpiling of global medical equipment ahead of future pandemics, serious reform of the World Health Organization and elevated funding into vaccine, antiviral and antibiotic treatments with the same intensity that we invest in weapons and space programs.

In a crisis, governments need to have consistent messaging from experts and be willing to impose politically unpopular punishments on individuals who put entire communities at risk by ignoring containment strategies.

Social media networks need to police public health misinformation on their sites. A Harvard Kennedy School study of 11,000 unique URLs, or web addresses, referring to the origin of COVID-19 appearing on Facebook, Twitter, Reddit, and 4chan in early 2020 showed that unproven conspiracy stories about the disease and its origins were far more viral online than neutral or debunking stories.[22]

Perhaps the most crucial lesson, though, is this: If we let a new pathogen scale up and rifle through the world's major population centers, we've already lost. Once you're in the pandemic phase of infectious disease progression, you're really talking about the terms of surrender, limiting the losses the best you can and making calculated tradeoffs between saving lives and avoiding social and economic collapse.

That's why in the wake of the COVID-19 ordeal, many of the world's leading epidemiologists, technologists and public health professionals believe we need to replace our existing pandemic playbook with a radically proactive approach, one that takes the contest with potential pathogens directly into the microbial realm.

That means building a sophisticated, technology-laden biosurveillance system that monitors microbes at the molecular level before they pose an immediate threat to humanity. It means gaining a far deeper situational awareness of the world's most dangerous microbes with genetic sequencing, powerful computational tools, as well as artificial intelligence (AI) and machine learning applications.

Modern-day meteorologists use an array of satellites, reconnaissance aircraft, radar, ships, and buoys to track the wind dynamics, dissipation, pressure and evolving structure of storm systems thousands of miles away to assess their risk of developing into a Category 5 hurricane.

The world's biggest financial institutions invest more on AI programs to detect subtle signs of credit card fraud than governments do to better understand dangerous viruses and bacteria that, under the right conditions, could kill hundreds of millions of us.

Avoiding future COVID-19 catastrophes will require training the world's best technologies on the microbial universe to better protect ourselves. The more we know, the more time we have to intervene and possibly prevent the next infectious agent from running wild and destabilizing the world again. And much like defense spending has created spin-off technologies with real commercial potential and applications that benefit humanity, so too might our pandemic prevention research.

We need to deal with these fascinating life forms on our terms, not theirs. And a useful place to start might be the world's public transportation networks, where scientists are now discovering a strange parallel universe of microbial diversity and entirely new species. When it comes to microbes, the Age of Discovery has only just begun.

Chapter 2

Surveillance for a Pandemic Age

Parents might be annoyed, even slightly alarmed, to see their young daughter lick the surface of a steel pole in the gritty New York subway system. Christopher Mason was intrigued. The geneticist and professor of physiology and biophysics at Weill Cornell Medicine in New York wondered how exactly we interact with the microbial world when we touch a public surface. What microorganisms are we comingling with, what are these life forms doing and how are they doing it?

Eventually, those musings led to a remarkable scientific quest to create a molecular map of the world's busiest mass transit systems. Microbes are swarming unseen all around and inside of us in intricate ecosystems called microbiomes. In the average human microbiome, bacterial cells are roughly equal in number to human cells.[1] Some 36 percent of the active molecules in our bloodstream typically come from microbes.[2]

Yet despite microbes being the oldest, most abundant and diverse life forms on Earth, scientists have only been able to study roughly 1 percent of them in controlled laboratory settings by growing them in cultures and really figuring them out.

Now, next-generation genetic sequencing technologies, gene editing and more powerful computing tools are allowing biologists and geneticists to explore this micro frontier, or what some call microbial dark matter. Along the way, scientists have learned more about the size and scope of microbiomes in human, terrestrial, aquatic, atmospheric, and even outer space environments.

Mason reasoned that if you want to find a rich tapestry of microbial life, what better place to explore than a subway system in a megacity with one of the highest population densities on the planet? In New York City, roughly eight million people are packed into a landmass of about 469 square miles. Worldwide, some 55 percent of the global population now live in urban environments, a figure that's projected to reach 60 percent by 2030.[3] Transportation systems are a daily nexus point for billions of commuters and microbial life.

In the summer of 2013, Mason and his colleagues launched PathoMap, as in pathogen map, with the aim of creating the first ever microbial atlas of an urban environment. The group's small army of researchers and citizen scientists took swab samples from a variety of surfaces at 466 stations, spanning across 24 major subway lines of the New York City Metropolitan Transit Authority.

They did the same at four public parks and the Gowanus Canal in Brooklyn, one of the most polluted waterways in the country and designated a Superfund cleanup site by the U.S. Environmental Protection Agency. Samples were geotagged to mark location and photographed, then taken back to a lab for DNA sequencing, analysis and comparison with databases of established microbial life.

Remarkably, in a study published in 2015, Mason's group reported that 48 percent of the DNA sequenced in the study didn't match any known organisms in existing genomic databases.[4] What's more, different parts of the city had unique microbial signatures when it came to the diversity of life forms and the demographic mix of the human's passing through.

One subway station on the southern tip of Manhattan, flooded during Hurricane Sandy in 2012, still had genetic readings for bacteria common to marine and even frigid Antarctic environments. The toxic Gowanus waterway was home to hearty creatures that devour sulfate and others that produce methane, microbial denizens that thrived in brackish and toxic watery environments.

The bacterial dynamics at Penn Station, a major node in the New York region's train network and one of the busiest transit hubs in the Western Hemisphere, were in constant flux as commuters, each with their individualized microbial profiles, collided on a massive scale every day.

Most of the bacteria, viruses and archaea species found in New York were considered harmless or even beneficially symbiotic to human health. There were genetic traces of microbial life forms that showed up on the government's warning list of infectious agents such as *Staphylococcus aureus*, a potentially dangerous bacteria. However, this study wasn't designed to determine if they were alive and metabolically active.

What the project did reveal was that it was entirely feasible to build a molecular map of an urban microbiome, one that could be calibrated for disease surveillance, bioterrorism threat mitigation, new drug development and even urban design.

So Mason and his colleagues scaled up their ambitions and created the Metagenomics and Metadesign of the Subways and Urban Biomes (MetaSUB), an international consortium of geneticists, engineers, data analysts, designers and public health experts. The group spent three years collecting about 4,700 samples from mass transit systems in more than 60 cities, from Barcelona and Guangzhou to Moscow and Sao Paulo. In a study published in May, 2021, MetaSub researchers reported the discovery of some 10,000 previously unidentified bacteria and viruses.[5]

In addition, the findings underscored how urban microbiomes were ecologically unique from surrounding land, water and human microbial communities, perhaps based on greenspace, tourism dynamics and local waste management systems. That said, there was a subgroup of microbes, mostly skin bacteria, common in almost all the cities explored by researchers.

The project's survey team also found evidence of antimicrobial resistance from samples taken from surfaces and the air around the world. This refers to bacteria, viruses and other microorganisms that have built up immunity to medicines designed to prevent infections and disease in humans and other animals.

The excessive use of antibiotics in medicine and agricultural practices worldwide has triggered a global spread of multi-drug resistant microbes known colloquially as superbugs. New strains of *Mycobacterium tuberculosis* are particularly concerning. More than 700,000 die annually from antibiotic-resistant bacterial infections. On current trends, more people will succumb annually to infections from drug-resistant microbes than to cancer by 2050, according to a study commissioned by the United Kingdom.[6]

I caught up with Mason in an early morning video call. He had overslept and hurriedly logged on, wearing a white T-shirt and looking a touch blurry-eyed. Understandably so, given the night before he had been up late dealing with a family emergency: sadly, the death of a relative from respiratory failure and other complications related to COVID-19. "He basically needed a lung transfer and that wasn't going to happen," Mason said. "It was his time."

Mason grew up in America's heartland, in a household full of computers and what he describes as an academic vibe. His mother was an English professor at the University of Wisconsin-Parkside, and his father served as the Chief Information Officer at the motorcycle manufacturer Harley Davidson. His interest in genetics traces back to primary school, when he started to grasp that the molecular instructions for life were encoded in every cell and that great stories were awaiting those who could read that book of life.

Later on, he became fascinated by the fact that whenever you sequence part of the human genome from somewhere in your body or skin, you'd always find fragments that didn't map to anything human and were, in fact, derived from microbial DNA. The interaction between humans and microorganisms would later spark Mason's interest in space exploration and, more philosophically, the survival prospects of the human species, indeed all biological life.

Mason helped conceive the landmark Twins Study designed to assess the long-term effects of space travel on the human body.[7] One of the subjects was astronaut Scott Kelly, who starting in March, 2015, spent 340 days in orbit on the International Space Station in the longest-ever NASA mission. His genetically identical twin brother, Mark Kelly, himself a retired astronaut and currently a U.S. Senator, was the other.

The study explored what extended spaceflight does to human cognition, cardiovascular health, genomic changes as well as alterations in gut, skin and oral bacteria. One odd finding was that Kelly's telomeres, the molecular caps on the end of his chromosomes that normally shorten as we age, somehow grew longer during his space mission. The heavy doses of radiation astronauts receive in space, the equivalent of multiple chest X-rays a day, triggered tiny genetic mutations in Scott that may have slightly raised his long-term cancer risk.

It was one of the most intensive analyses of any individual human body in history and one full of epic logistical challenges. The explosion of

Elon Musk's SpaceX Falcon 9 rocket in 2015 soon after takeoff delayed one experiment-related, incoming supply mission bound for the space station. Getting Kelly's blood, saliva, urine and stool samples back down to Earth required hitchhiking on departing cargo spacecraft, whose capsules typically landed in Southern Kazakhstan, then had to be airshipped to Houston.

Life on Mars

Mason thinks our future survival depends on learning from the microbes, or at least adapting some of their molecular features to tolerate the extreme environs of outer space and other planets. He wrote a book about his 500-year plan to get there.[8] This isn't just an interesting thought experiment. Scientists believe that the Earth will become uninhabitable for complex aerobic life like us in about one billion years.

As our sun ages, it will become hotter and release more energy. The destruction of photosynthesising organisms will deplete oxygen and, with it, destroy the atmosphere's protective ozone layer. That in turn will eventually open the Earth and its oceans to overwhelming levels of heat and ultraviolet radiation that will eventually wipe out most terrestrial and aquatic life. Life on the planet will be primarily microbial, just as it was billions of years ago. Further out and five billion or so of years later, the sun will run out of hydrogen, start burning helium, enlarge into a red giant and likely incinerate all biological life on Earth.

Disheartening to contemplate, yet Mason is an optimist and thinks science will give us options to find new homes before then. To get there, though, we'll need to achieve a scientific and cultural consensus to radically reengineer the human genome for the long-term survival of the species. "There are things to worry about when it comes to modifying species, modifying ecosystems," explained Mason. "Yet if we don't do it, we run the risk of all life being lost at some point."

He believes that the class of microbes called extremophiles, hearty microorganisms that survive in extreme temperatures or in acidic conditions, have much to show us about surviving long-term in space as we set our sights on colonizing Mars mid-century. By 2300, as Mason sees it, we may be sending microbial genomes to Earth-like planets to seed life.

A more immediate concern is making sure a dangerous pathogen doesn't kill all of us in the here and now. Mason thinks we have a lot of technology already in place to make big strides in our pandemic surveillance. MetaSub and similar programs are proof of that concept. You could easily refocus such efforts on reading the subtle molecular changes of disease-causing pathogens and take measures to contain them before they spread aggressively in human populations.

"For the longest time, people thought it would be too laborious, or too expensive, to go pathogen hunting for zoonotic sources in urban environments," Mason said. Post-COVID 19, "people now recognize the cost of not doing it is in the trillions of dollars of lost economic growth and productivity."

With the price of genetic sequencing continuing to plummet, a worldwide microbe sentinel system is feasible, though it would probably also require tracking these infinitesimal life forms on the land, sea, air and outer space. "People pay taxes for a military that defends our national security. Pathogens are in a sense our enemy," Mason argued. "You could do a lot with one billion dollars." The estimated cost of dealing with the COVID-19 pandemic is expected to hit $22 trillion by 2025.[9]

Investing in a robust pandemic surveillance system might also have potential upsides in new drug development and commercial spinoffs, according to Mason. Most of the viruses captured by the researchers in the MetaSub study were bacteriophages, better known simply as phages, one of the biosphere's most abundant organisms that engage in non-stop chemical warfare with bacteria.

Understanding better how bacteria evade infection, paradoxically, could result in a deep pipeline of new and more powerful antibiotic drugs. "The poetry of the data is that it's both a quantification of our risk, but also the source of new tools to fight that risk," according to the geneticist.

In their vast interactions with each other, microbes produce a multitude of spin-off, biochemical processes called secondary metabolites. They're secondary because these features aren't directly associated with the growth of a microbe, but often provide a key selective advantage such as cleverly warding off a rival bacteria or producing steroids, scents and oils. These antibiotic byproducts from microorganisms have been very useful as well in curing human diseases caused by bacteria, fungi and protozoa.

In 1928, in one of the most remarkable chance discoveries in scientific history, bacteriologist Alexander Fleming noticed that a mold growing in a discarded petri dish had killed a bacteria called *Staphylococcus aureus*. The active antibiotic agent in the mold turned out to be penicillin, a revelation that later set off an astounding period of antibiotic discoveries and medical advancements that saved hundreds of millions of lives in the twentieth century.

Mason's team identified molecular formations, called biosynthetic gene clusters, that may set the stage for the production of these secondary metabolites. In addition, insights into the adaptive immune systems of bacteria as they protect themselves from viruses also gave Mason's team ideas for drug development using CRISPR gene-editing technology.

As we move to protect ourselves against pathogens, we may simultaneously usher in a new era of "natural" drug development. "There are a lot of data, better computational tools and high through-put ways to make cells or modified cellular systems that were difficult to do 5 or 10 years ago," according to Mason.

Virus Sleuthing

The search for the next pandemic-spawning microbes extends beyond urban centers. Many of the world's emerging disease hotspots are in remote locales, where wildlife, livestock and humans intermingle in potentially dangerous ways.

Any robust biosurveillance system will need to focus on outlying areas where forests have been razed for agricultural, residential and business development; where communities engage in economic and cultural practices that enhance their risks of encountering pathogenic microbes; and where wildlife and bushmeat markets and supply chains capture and consume virus-carrying animals.

Testing species suspected of having high viral loads such as bats, monkeys and rats can lead to the discovery of new infectious agents that haven't yet burst onto the world stage. That's a much better strategy than reactively responding to a runaway outbreak with blunt tools like massive quarantines, international travel bans and wealth-destroying economic lockdowns.

Taking pathogen surveillance to the frontlines of emergent infectious disease circulation in nature is critical if we're serious about lowering pandemic risk, according to Dennis Carroll, former director of the U.S. Agency for International Development's (USAID) Pandemic Influenza and other Emerging Threats Unit. "We are still stuck in a very twentieth-century mode of finding a pathogen, developing a vaccine against that pathogen, and waiting for the next pathogen and developing a vaccine against that pathogen," he said.

Carroll earned his doctorate in biomedical research, with a focus on tropical infectious diseases, at the University of Massachusetts Amherst and worked as a research scientist alongside James Watson, the co-discoverer of DNA with Francis Crick, at the Cold Spring Harbor Laboratory on Long Island, N.Y., in the 1980s.

Before he left government work, which also included a stint at the U.S. Centers for Disease Control and Prevention (CDC), Carroll and other top infectious disease experts in 2016 announced plans to create an atlas of the world's assemblage of viruses to better understand their transmission dynamics and provide timely data for public health interventions before outbreaks turned into epidemics and pandemics. They called it the Global Virome Project.

With his shoulder-length white hair and fondness for turquoise jewelry and Asian art, Carroll gives off a slightly bohemian air. During stretches of the COVID-19 pandemic, he sheltered in a houseboat on the Potomac River in the Washington, D.C. area.

Earlier in his career at the CDC, he oversaw the agency's portfolio of programs in malaria, tuberculosis, antimicrobial resistance, and disease surveillance. Carroll led efforts to deliver treatments for onchocerciasis, better known as "river blindness," the development of rapid diagnostics for malaria and treatments against drug-resistant versions of that disease, as well as the creation of insecticide-treated bed nets. At USAID, he also researched lethal, airborne avian influenza strains that with the right mutations could kill millions of us.

Carroll wasn't all that surprised that a coronavirus triggered a devastating pandemic. Over the past 20 years, there have been two other deadly coronavirus outbreaks: severe acute respiratory syndrome (SARS) in 2003 and the Middle East respiratory syndrome (MERS) in 2012.

The viral threats haven't stopped there. There have been serious influenza flare-ups in 2003, 2009, and 2013. Since its resurgence in 2013, Ebola, a pernicious hemorrhagic fever disease, has been a consistent challenge in West and Central Africa. The Zika virus, first discovered in 1947 in the Zika Forest of Uganda, swept through Brazil and other Latin American countries in 2015 and 2016.

Many of these disease-causing viruses are zoonotic, meaning that they somehow jumped from wildlife or livestock reservoirs into the human population. There may be an additional nearly 1.7 million yet-to-be-discovered viruses in 25 viral families known to exist in mammal and bird hosts, and anywhere between 631,000 to 827,000 of them may have the capacity to migrate across species, the Carroll team's research has shown.[10]

The twenty-first century has seen an acceleration in epidemics and pandemics, powered by urbanization, globalized trade and travel, environmental degradation and climate change. If you want to calibrate the risk of a geographic region becoming an infectious disease hotspot, check out whether there's been a dramatic change in its land use that's brought humanity in closer contact with wildlife reservoirs or livestock. "It's the number one predictor of trouble," says Carroll.

While at USAID, Carroll teamed up with Jonna Mazet, a professor of epidemiology and disease ecology at the University of California, Davis, School of Veterinary Medicine. She led a group that had designed a disease surveillance program called PREDICT that eventually spanned 35 countries and built diagnostic systems to identify new viruses in regions believed to be high-risk for infectious disease outbreaks.

Mazet's infectious disease hunters focused both on the structural features of viruses, such as their binding properties to human cells, but also their elasticity or ability to migrate to different species of hosts. Not every virus is a threat. The aim is to figure out which viruses to really pay attention to.

In 2018, PREDICT researchers discovered and genetically sequenced entirely new species of the Ebola viral family called *Bombali ebolavirus* and found the super-lethal Marburg virus in bats in Sierra Leone in West Africa.[11] This kind of early viral discovery is crucial to assessing the risk of pathogens jumping species and causing new outbreaks.

The other important epidemiological detective work involved connecting animal reservoir hosts to the various strains of the Ebola virus that have been killing Africans since the late 1970s. Knowing which animals in your community pose infection risk can save lives. Mazet's scientists detected the Zaire variant of Ebola in a bat in Liberia, but it's still unclear whether the mammal is the original host for the virus. The Zaire Ebola virus killed more than 10,000 people between 2013 and 2016.

PREDICT and its research partners have done plenty of work on coronaviruses in Asia, Africa and Latin America and found about 160 new species then unknown to science. Unfortunately, SARS-CoV-2 which causes COVID-19 went undetected and by the time the world caught up with that pathogen it was too late. A deeper pool of genetically sequenced coronaviruses, maps of identified animal carriers and detailed knowledge of the supply chains within the wildlife markets in China, might have made a difference.

"The COVID-19 virus was probably a real, active virus for years, but it was off in a bat somewhere," said Carroll. "Nobody really wanted to pay attention to that because it wasn't a clear and present danger. We need to reengineer funding to think about risk as a valid space for investment."

Ask Carroll which future viral threats he worries most about, and he'll point to two. One is H5N1, a highly infectious and deadly respiratory virus in birds that's occasionally spilled over into humans and killed 60 percent of them.[12] Right now, it doesn't transmit efficiently to humans, but it's constantly evolving. "The Egyptian variant is three mutations away from what laboratory work has suggested would be an efficient human-to-human transmitter," Carroll said.

The other is the Nipah virus, which kills up to 75 percent of those it infects with symptoms ranging from muscle pain and sore throats to encephalitis, or brain inflammation, seizures, and coma before death.[13] It tends to spread from fruit bats during date palm sap harvesting season in Bangladesh and other parts of Asia. ("Bats have sweet tooths," Carroll noted.) It can be transmitted human-to-human but hasn't yet broken out on a major scale.

"It goes four or five people deep, then it dies out. It doesn't yet have sustained transmission potential," said Carroll. "Every year, the virus is looking for that golden mutation that will enable it to be sustained human-to-human." The killer disease in the 2011 thriller *Contagion* was modeled on a super-strain of Nipah imagined by several PREDICT scientists who served as technical advisers on the film.

Researchers have also deepened their understanding of how economic practices in less-developed parts of the world result in risky microbial exchanges. In Myanmar and Cambodia, farmers and suppliers supplement their income in those regions by exploring bat caves in search of excrement to fertilize crops and gardens, placing them at high-risk of infection. The hunting of bats for consumption and trade of their meat is also common in West Africa. All of this despite the fact that many bats are well-established reservoir hosts for viruses that cause rabies, Ebola, SARS and MERS.

By using artificial intelligence software models of the viral dynamics between animals, PREDICT researchers developed data that showed that bats, followed by rodents and shrews, were at higher risk of shedding viruses compared to other wild mammals.

There were also seasonal patterns to infectious disease risk: Viral microbes from the major high-risk pathogenic groups (coronaviruses, flaviviruses, paramyxoviruses and influenza viruses) were more likely to be found in animals during wet seasons. Live wild mammals transported in supply chains and sold at markets had significantly higher viral loads, given their close proximity to each other in cages and greater opportunity for microbial exchange from excrement, bodily fluids and respiratory droplets. That kind of on-the-ground sleuthing about animal transmission routes is essential to prevent epidemics that grow out of control, according to Mazet, the former global director of PREDICT.

"If you know what the common animal carrier is, you can shut down these outbreaks at the source," she said. By better understanding and mapping the behavior of bat species in West Africa, local health officials started to get a more nuanced view of Ebola transmission risks. In that part of the world, "bats were going in and out of houses" but few understood the dangers of that kind of close contact, she said.

Mazet sees an urgent need to set up a global, genomic-based biosurveillance platform that would intercept microbial threats before they evolve into costly and lethal epidemics and pandemics. The Global Virome project, of which she's a board member, would be one pillar of that effort.

The other two would be research programs such as the International Barcode of Life Consortium BIOSCAN and the Earth BioGenome Project that are trying to sequence, catalog and characterize the genomes of complex living organisms on the planet.

If we devoted a fraction of what we've spent fighting the COVID-19 pandemic, some infectious disease experts argue that we could make rapid progress identifying the pathogens hosted by mammals and birds and have a shot at avoiding another costly global health emergency. "We are going to need more data-sharing agreements and data-sharing platforms," Mazet said.

She points to one promising interactive web tool already up and running called SpillOver, which allows local health authorities and researchers to calibrate risk for potential wildlife-origin viruses in their regions. Leveraging data from more than 500,000 genetic samples from nearly 75,000 animals, plus public records of viral detections worldwide, SpillOver calibrates risks for about 890 viruses.[14] It's an open-source platform and researchers worldwide are now uploading new data to refine the model.

Since leaving U.S. government work, Carroll has continued to lobby for international funding for his virome project and better pandemic surveillance. He has enlisted such notables as former British Prime Minister David Cameron to help drum up support from the Group of Seven heads of states.

Given the plunging costs of genetic sequencing and computing power, Carroll is convinced that his project, with an operating budget of about $100 million per year, could over the next decade identify the world's biggest microbial threats, build a comprehensive database of their key biological and molecular traits, and design a robust surveillance and preparedness system for governments and scientists to draw upon.

Even given the trillions of dollars lost from the COVID-19 calamity, Carroll thinks raising money won't be an easy sell. Governments have a hard time selling voters on funding abstract public health risks that may be in the distant future. "Money moves against known threats," he said. "Money has a hard time moving against the risk of a threat."

Mosquito Mercenaries

One step removed from the pathogenic microbes are the animal and insect vectors that transmit them. Disrupting those channels is another emerging strategy in pandemic preparedness.

In May, 2021, a British biotech firm spun out from Oxford University called Oxitec made a bit of scientific history by releasing genetically modified mosquitoes into the air over the Florida Keys. The pilot program, the first of its kind in the United States, shows how advances in genetic engineering may play a role in pandemic prevention. Oxitec spent about a decade painstakingly working through the U.S. regulatory process and educating the public.

Now owned by U.S.-based Third Security, Oxitec has developed a way to suppress the population of an invasive species of mosquito called *Aedes aegypti,* which originated in Africa and has spread far and wide in a more globalized world. The *Aedes* species makes up just 4 percent of the mosquitoes in the Keys, but transmits nearly all of the Zika virus, which can be passed from pregnant mother to child, causing severe fetal brain defects.

This insect has adapted well to urban areas, has a fondness for human blood and is a day-biting mosquito; such habits have reduced the effectiveness of sleeping nets. One study forecasts that by 2050, some 49 percent of the global population will be exposed to *Aedes aegypti* mosquitoes, making these insects one of the most concerning disease vectors in the biological world.[15]

Aside from Zika, the *Aedes* species has also powered a thirty-fold increase in dengue fever, which causes a severe and sometimes deadly flu-like illness. Dengue doesn't spread directly person-to-person. However, infected patients can transmit to virus-free mosquitoes that bite them, which then in turn spread the pathogen to new human hosts. This species also spreads maladies such as chikungunya and yellow fevers.

Founded in 2002 by two Oxford University zoologists, Luke Alphey and David Kelly, Oxitec has developed genetic tools designed to crash the population of *Aedes* mosquitoes and by extension suppress the spread of diseases they carry.

It does so by genetically modifying male *Aedes,* so that when they mate with females their offspring inherit a gene that overproduces a protein. It's lethal to females, which are the biters in search of blood to produce eggs; male descendants aren't affected but pass along the life-shortening gene to other females as they mate. Eventually, the population starts to collapse. Oxitec has reported that its trials in Brazil and Panama have shown population reductions of more than 90 percent.

A second addition is a fluorescent marker gene that produces another chemical that glows when exposed to a specific frequency of light. That allows government pest control authorities to track the insects in the wild and monitor progress.

Figuring out ways to disrupt viral transmission networks is a promising area of disease prevention, if it can be done safely. Activist groups and academic critics have pointed to the risk of unintended consequences from such genetically engineered interventions. What if, for example, a new super-strain of mosquitoes accidentally emerged making the problem worse?

Oxitec Chief Executive Officer Grey Frandsen thinks critics of genetically modified organisms are wrong on the science and have used unsubstantiated claims and scare tactics that generate a lot of headlines but do little to advance public health. With government approvals and endorsements from a range of countries, universities, institutions, foundations, and communities, Oxitec has released one billion of its altered mosquitoes worldwide in trials with no mishaps and have documented results showing its effectiveness, Frandsen said.

Governments, meanwhile, are scrambling to find new tools to combat this disease vector. Zika and dengue are serious public health threats, and malaria is making a comeback as climate change widens the reach of disease-carrying mosquitoes that are building up resistance to insecticides.

"Mosquitoes have been able to out-engineer our historical attempts to control them," explained Frandsen, whose career in government work with the U.S. State Department, Centers for Disease Control and Prevention, and biotech startups has taken him from posts in the Balkans to Africa, giving him a broad perpective on the global health challenges we face in the twenty-first century.

"We're focused on the vector of these devastating diseases. Our technology has the ability to suppress targeted mosquito populations and to reverse insecticide resistance in remaining local populations, all without leaving a trace in the environment," he said in an interview. "This clean, targeted technology is precisely the type of tool that will be needed as traditional approaches lose their effectiveness."

Oxitec is also in the progress of launching another pilot program in California, where *Aedes aegypti* mosquitoes have made big inroads in recent years. "Climate change is rapidly expanding the threat that disease-transmitting mosquitoes pose to public health," according to Frandsen. "This means that millions more people are now under threat of deadly or devastating diseases."

Oxitec still needs to win U.S. regulatory approval to sell its technology on a commercial basis. Frandsen is confident that will happen and thinks vector suppression will be critical in pushing back the global resurgence of malaria, a disease that he has contracted twice while working in Africa.

Oxitec is now developing two new mosquitoes, one of which targets *Anopheles stephensi.* Malarial vectors have tended to be prominent in rural areas, but this species is taking the disease to urban environs, according to Frandsen. "Currently there are no effective tools to challenge this vector and it could soon impact nearly two billion people."

This emerging malarial threat points to another, broader issue for public officials trying to deal with infectious disease spread. We have little, real-time, situational awareness of where the pathogens are and where they might be heading next.

"We know new public health threats are coming, and they could emerge from anywhere. Our challenge now is to build a global surveillance system that can detect, share, and monitor the health threat. That requires each country to strengthen surveillance systems and view itself as a critical node in a global network that works seamlessly," Frandsen said. "How quickly can we digest data and in one pin-point location, and how quickly do we share that data with a global coalition, will mean the difference between an isolated breakout and global pandemic. The future of our species on this planet will be deterimined by how effective this network is."

Pandemic surveillance is all about finding potential microbial threats and disrupting them before they have a chance to scale up. Another powerful tool in this effort will come from artificial intelligence technology that can quickly see molecular patterns in viruses and bacterial microbes that humans can't. Call it the bots versus the bugs.

Chapter 3

Hacking into Nature's Secrets

Back in 2003, as a novel coronavirus that caused severe acute respiratory syndrome (SARS) stormed out of Southern China and spread to 29 countries, Kamran Khan had no idea what was heading his way. A physician, infectious disease expert, and professor of medicine at the University of Toronto, Khan's colleagues suddenly contracted the lethal virus and front-line healthcare workers in the Canadian business and financial center lost their lives after a local outbreak.

In all, about 44 Canadians were swept away by the global contagion of SARS, which packed far more lethality than the more recent SARS-CoV-2 virus that has caused COVID-19 and destabilized the world since late 2019. The overall case-fatality rate for SARS was 9.6 percent in Canada and 45 percent for infected patients older than 60.[1] In contrast, it has been less than 1.7 percent during the COVID-19 pandemic.[2]

Fortunately, that SARS virus nearly two decades ago lost its transmission momentum and since 2004 there haven't been any known cases of it reported in the world. Yet an infectious disease specialist like Khan realized how vulnerable the world had become to fast-moving pathogens, and in 2013 he launched BlueDot, a data research and digital health firm focused on worldwide infectious disease threats.

The company has developed a comprehensive epidemic intelligence system based on artificial intelligence (AI) and a multi-disciplinary team of computer scientists, climate specialists, microbiologists, and data crunchers.

BlueDot's AI-guided Global Intelligence Platform scans hundreds of thousands of different sources, including online articles in nearly 100 languages, looking for early signals of outbreaks arising from more than 150 different diseases. It does so every 15 minutes, 24 hours a day.

In 2016, the company predicted the arrival of the Zika virus to Southern Florida from Brazil, six months before the first reported case in Miami. It did so by analyzing travel patterns and the climate conditions for mosquito populations that transmit the virus.

The firm's surveillance system uses natural language processing and machine learning to filter, contextualize and extract the information that truly matters. Data gleaned from an array of online open-sources can supplement traditional data collection and analysis from official ones, but they can't entirely supplant human expertise.

Getting the search parameters right matters. Back in the early 2010s, Google rolled out a Flu Tracker that promised to "nowcast" influenza seasons by analyzing search terms. The tech giant quietly shelved the product after missing an earlier flu surge by a wide margin. An autopsy study of the flop called it an example of "Big Data hubris."[3]

BlueDot's system is more advanced and is backed up by Khan's staff expertise. On December 31, 2019, the company's surveillance system picked up and translated a Chinese-language news report of a mysterious pneumonia in Wuhan that wasn't caused by usual microbes like the influenza virus. BlueDot's team quickly understood the deeper significance, and within two hours, the company issued a bulletin to its government and private sector clients warning of possible trouble.

Next, BlueDot's data analysts started to look at aggregated travel and mobile phone data in and around Wuhan, a megacity of 11 million and the capital of Hubei Province in central China. By extrapolating from recent travel patterns, it became apparent that there was a great deal of mobility between Wuhan and Seoul, Tokyo, Shanghai, Hong Kong, Shenzhen, Bangkok, Taipei, San Francisco and New York. BlueDot's system accurately identified the emerging contours of a global contagion earlier than most.

Spend any time with Khan, and a couple of things become clear. First, while he's a big believer in the power of Big Data analytics and AI applications to better detect emerging disease threats, he's no technology evangelist. "AI isn't some quasi-mythical creature that can just come up with these insights,"

Khan told me. "Let's not get enamoured with technology alone. Doing better is also about human factors, not just machine learning."

In his view, the faltering global response to the COVID-19 pandemic is an innately human and societal problem, starting with our inability to assess long-term, infrequent threats like epidemics until they're tearing societies apart. "The reality is that our brains are not wired for slow moving threats and we discount the significance of future events," Khan explains. "It's impossible to plan the escape route, when the house is already on fire."

More optimistically, Khan believes there is a chance that the COVID-19 pandemic has ushered in a global awakening of sorts about the growing risks from infectious disease outbreaks. Some six million in confirmed global deaths and trillions of dollars in economic pain have prompted the U.S., European and Asian governments to grasp the urgency of pandemic preparedness.

"I think what will ultimately happen is that the cadence of these types of emergencies, not necessarily on the COVID scale, is on the order of every few years," reckons Khan. "So we will be getting more stimuli coming forward to remind us that the problem hasn't gone away."

For Khan, microbial disease threats are heterogeneous, complex and fast-moving phenomena, requiring a data and analytics platform that can deliver timely and reliable information to multiple actors to pull off the kind of coordinated response that can make a difference in managing the spread of viral and bacterial pathogens.

Disease outbreaks are societal problems requiring a societal response, according to Khan. Relying on a handful of government and health officials to decide what and when to disclose time-sensitive information to the public is a recipe for failure.

Corporate risk-managers, point-of-care healthcare workers, educators and ordinary citizens need to be plugged into this information ecosystem. Viruses are fast but so are emails and text message alerts. "If we really want to have this kind of resilience, we are going to need to be empowering a whole bunch of different audiences," he said.

In Khan's assessment, we now identify and share information about pandemic threats in an ineffectual, trickle-down system that cascades from the government to the healthcare system and industry, and then only belatedly to the general public. That glacial pace cedes huge advantages to a highly contagious microbe.

One person armed with the right information can have an outsized impact on the course of an epidemic, particularly front-line healthcare workers. "This is my specialty and it's hard for me to stay on top of what diseases might be spreading in some remote jungle somewhere in Borneo, let alone a family physician who doesn't specialize in this area," Khan said.

Fighting disease threats also needs to be crowdsourced to the public to some degree. The act of reporting illnesses and symptoms that seem out of the ordinary after traveling needs to be as reflexively obvious as calling emergency services to report a house fire.

Governments have big disincentives to draw immediate attention to emerging disease threats. They threaten powerful corporate and economic interests and can upend political careers. The same may hold true for the World Health Organization, whose more powerful member states like the United States and China potentially have a lot to lose economically and geopolitically from decisions made in Geneva to declare global health emergencies or pandemics.

On average, BlueDot picks up reports of outbreaks, most of which are localized, about 40 days before government health authorities. That underscores the need for independent, third-party public health information, free of political considerations.

The resilient public pandemic surveillance network that Khan has in mind, perhaps unsurprisingly, is similar to the one he built at BlueDot, which is focused on three core tasks: detecting, assessing and responding to infectious disease threats.

When it comes to detection capabilities, BlueDot's Global Intelligence Platform scans both official sources like government and public health reports, as well as informal ones like media posts. The information is sometimes fragmented and unstructured, and the company's algorithms are looking for signals in the noise.

Khan sees a big opportunity for governments and healthcare systems to invest in better syndromic surveillance networks. Are people suddenly searching for information about a particular type of illness? Are emergency departments picking up an increase in a certain type of respiratory or diarrheal illness?

Imagine a digital alert system that informed healthcare workers of urgent health trends at the grass-roots level. One well-informed emergency room physician could potentially prevent a novel disease from ever turning into a pandemic.

Much more data about pathogenic microbes should be shared with front-row clinicians, Khan argues. "We need better diagnostic capabilities, so that we can rapidly at the point of care identify a particular microbe," he said. In Canada, samples from an unusual disease can take weeks to be analyzed at national labs.

A 2021 study led by Madhukar Pai, a physician and professor of epidemiology at McGill University, showed that systemic lack of investment means basic diagnostic capacity is available at only 1 percent of primary clinics in many low- and middle-income countries.[4] "If these countries do not have the tools available to diagnose the common cause of fever or cough, how can they spot potential outbreaks?" the study concluded.

Khan sees value in scientific research to map out and sequence more of the microbial world, though he sees challenges in identifying the novel viruses that actually have the potential to infect humans. "That's part of the long play: basically cataloging what's out there," said Khan. "The reality is that we know so little about our microbial environment."

Once a disease threat has been detected, BlueDot's team focuses on two sets of risks: dispersion and disruption. If humans are the primary transmission vector, then tracking population movement via smartphones and other digital devices can be useful.

"When we picked up this early signal of an outbreak in Wuhan, within about two seconds our system had talked to all of the anonymized mobile phone location data, looked at all the neighboring airports, the probability the movements would happen to which airport, to which flights it was connected to, and passenger movements," Khan points out. "We knew how the (ground-zero, seafood) market in Wuhan, at that moment, was connected to the rest of the planet."

Not every new virus or bacterium turns into an epidemic; even fewer morph into pandemics. So the details matter. What's the community spread risk, climate conditions, availability of animal and insect transmission vectors, and the quality of the local health system? Is the outbreak going to burn out? Or do its embers have the potential to trigger a wider conflagration?

The response piece of the information ecosystem that Khan envisions can run the gamut from greater public awareness and routine health precautions to flight restrictions and national lockdowns. Yet the more informed various stakeholders in a country, the more likely they'll be willing to make the necessary sacrifices to halt the community spread of a novel disease.

Having independent, crowd-sourcing data firms out there can be a good check against politicized healthcare policies in the world's government capitals. That said, no surveillance system is full-proof against invisible, mutating microbes, even one powered by powerful AI technology, Khan concedes.

The last several years have been humbling, both in our inability to contain SARS-CoV-2, but also in what the pandemic has revealed about our cognitive biases and government funding priorities, according to Khan.

"At the height of the pandemic, we were having a U.S.-style 9/11 every single day in terms of death," Khan muses. "Looking forward, would we expect it (more) likely that a country would face a land invasion leading to millions of deaths? Or will it be an insidious virus that slips across the border into the country? Which is more probable?"

Siri, What's Ebola?

In the 2020s, AI disciplines like machine learning, natural language processing and big data analytics can give us an edge in our evolutionary struggle with pathogenic microbes, particularly when it comes to early infectious disease detection, rapid-fire diagnostics and drug development.

The technology is nowhere near good enough yet to take human expertise out of the equation entirely. It is nevertheless powerful and offers a useful weapon for microbiologists, virologists and other infectious disease experts trying to give us a fighting chance in preventing the next big pandemic.

To fully realize AI's potential, we'll need to rethink and deepen our data of the microbial world, figure out ways to standardize that information, and create global platforms accessible across scientific and medical disciplines, particularly the front-line clinicians. Right now, the data stream is chaotic and disjointed.

"Nobody has all the data they need in one place," Nita Madhav related to me in an interview. She's the Chief Executive Officer of Metabiota, a San Francisco-based digital epidemiology data firm and consultancy that develops predictive models using AI, statistical analysis and data tools for insurance companies, governments and other clients. "We'll need a far more collaborative approach," according to Madhav.

And while there's a robust debate to be had about guaranteeing the privacy of individual patient data, we'll need to be more open to using anonymous metadata that tracks health trends across a population for clues about how a new pathogen is evolving and spreading.

The COVID-19 shock has certainly made political leaders and individuals more aware of microbial threats. Getting the public more comfortable with opening up their medical histories for the greater public good is entirely another matter.

AI is a set of computational and analytical technologies that are widely discussed but thinly understood. Its theoretical roots date back to the mid-1930s, when the British logician and computer developer Alan Turing explored the possibility of a machine modifying, or even improving, its own program of instructions.

The actual phrase "artificial intelligence" was given life at a Dartmouth University summer research gathering by a small group of academics in 1956. The mission of this group, as described by Dartmouth mathematics professor John McCarthy, was "to proceed on the basis of the conjecture that every aspect of learning or any other feature of intelligence can in principle be so precisely described as a machine and be made to simulate it."[5]

Since then, the perceived promises and perils of AI have become a cultural trope, from Stanley Kubrick's late-1960s vision of the creepy heuristically programmed algorithmic computer (HAL) in *2001: A Space Odyssey* to the various *Terminator* films about time-traveling cyborg assassins.

Yet if you're a geneticist or microbiologist interested in understanding the intricacies of microorganisms that might wipe out hundreds of millions of us, AI doesn't look all that threatening. Rather, it's proving to be a rather powerful investigative tool.

AI has many subdisciplines such as machine learning, data mining, natural language processing, robotics and computer vision, though there's a fair amount of overlap across these specialties. For microbiologists and epidemiologists, machine learning applications are starting to be useful in helping decipher the nuances of microbes and solving other biological problems.

Machine learning programs use algorithms, or a set of computational instructions, to find discernible patterns in a vast miasma of data. That includes "structured" data like names, addresses, credit card numbers and geolocation, but also "unstructured" information like social media, business documents, images, video, audio and live chats. To understand the latter, requires the ability to sort, refine, deduplicate, categorize and understand context and linguistic nuance.

The healthcare world is teeming with unstructured data pouring out of medical devices like endoscopes, surgery robots, magnetic resonance imagers and emergency treatment video cameras. Then there's the biosignal information from patient monitors and wearable health monitoring devices, or the back and forth between doctor and patient chronicled in text notes or audio files.

A subset of machine learning is called deep learning, which very roughly mirrors our brain's neural network, in the sense that many layers of computational mini-networks collaborate in analyzing data and making predictions. Virtual assistants like Siri and Alexa rely on deep learning to understand user requests. Self-driving car developers such as Google and Apple use this neural-like technology, as do entertainment purveyors such as Netflix and Spotify.

Deep learning is emerging as a critical factor in medical diagnostics and research, helping physicians diagnose diseases and pathology results, and better understand and synthesize the genetics of microorganisms.

Roughly five billion people around the world have mobile phones, many of which are smartphones with powerful, high-resolution cameras, apps and wireless connectivity that can send and receive vast amounts of data. That presents opportunities for real-time disease tracking and diagnostics. Healthcare entrepreneurs and technologists are coming up with some innovations that could change twenty-first-century medicine.

Front-line healthcare workers now have access to sequencers that are mobile and can be carried in a small suitcase to decode the genomes of dangerous viruses and bacteria. A company called Oxford Nanopore Technologies, an entrepreneurial spinout from the University of Oxford, has developed a handheld DNA and RNA sequencing device called the MinION that costs roughly $1,000 and runs off a laptop USB plug.

"You can sequence 12 to 24 samples in a day, and you get a good estimate of the whole genome," according to Andrew Rambaut, a professor of molecular evolution and bioinformatics at the University of Edinburgh. He has trained healthcare workers in Central Africa to use such technologies to trace microbial mutations.[6]

Research has shown that attaching magnifying devices to high-end smartphone digital cameras could provide an affordable way for infectious disease clinical care in remote areas. AI-powered image recognition software is being used to diagnose diseases or assess symptoms, though it's still early days and human experts still need to monitor the results.

Smartphone apps have also played a useful role in helping scientists understand the dynamics of the mutating SARS-CoV-2 and diagnose infections. A COVID-19 symptoms research app that runs on Android and iOS developed by King's College London, Guy's and St. Thomas' NHS Foundation Trust hospitals and a health science company called ZOE Global has attracted more than four million volunteers globally, making it one of the world's largest ongoing studies.[7]

By analyzing reports of COVID symptoms with AI-powered algorithms, the project has been able to track viral transmissions and hotspots in the United Kingdom and elsewhere. The data also helped create a predictive model that showed that anosmia, or the loss of the sense of smell, was a key symptomatic indicator for testing positive for COVID-19.

Citizen scientists are also helping researchers capture disease carrying mosquitoes for analysis. A real-time mosquito surveillance program called Mozzie Monitors is a crowd-sourced program started in Australia in 2018, in which individuals set up mosquito traps in their backyards and use an app to monitor results.[8] Being able to track urban mosquito migration patterns affordably has been beneficial to public health.

Things might really get interesting if researchers and entrepreneurs can refine strategies to develop biosensors that can talk electronically to a smartphone, allowing it to recognize a targeted molecule in a biological sample. Researchers at Newcastle University in Northeast England have developed an ultra-thin protein layer, about 30,000 times thinner than a piece of A-4 paper, that shows promise. It's designed to trap proteins from disease-causing viruses that will change the electrical characteristics of the layer.

A university spin-off company called Orla Protein Technologies Ltd has fashioned a customized chip that can read the information from the protein layer. If blood and saliva samples collected in a small device with the protein paper and a biochip test positive for a disease in question, a signal is sent to a smartphone and onto a lab.[9]

During the pandemic, there was an explosion of more than 20 apps and other digital systems for contact tracing of those infected or at risk of being so, prompting one publication to create an app to trace the world's contact-tracing apps.

Some were very intrusive. China not only used location data to monitor the quarantined and track contacts, but also payment data. Beijing deployed CCTV cameras pointed at the apartments of those ordered to undergo 14-day quarantines; drones were used to order people to wear masks.

Israeli security agency Shin Bet, which has tracked cell phone data for its counter-terrorism efforts, used the same tactics to fight the spread of COVID-19.[10] In Singapore, the government-issued app used Bluetooth signals between mobile phones to track whom potential carriers of the disease had been in contact with. The disease-fighting potential of smartphones married with AI and data analytics may be too compelling to ignore, and it's likely these kinds of innovations will continue as we prepare for the next global outbreak.

There's some evidence, based on the success of the COVID Symptom Study in the U.K. or Australia's mosquito surveillance program, that individuals are willing to participate if there's a clear rationale for the research and reasonable privacy safeguards. Yet in some countries where government distrust is high, it will take a sustained public education campaign to win hearts and minds.

Catastrophic Risk

Big data analytics and AI can also help businesses and insurers better assess financial risk from microbial threats. Infectious diseases and associated mortality ebb and flow. New pathogens can die out quickly. Other long-lasting endemic scourges like tuberculosis and malaria can flare up.

It's not just the novel viruses that we haven't previously seen that risk analysts fret about. The vulnerabilities we know about are quite worrisome as well. A serious influenza outbreak would cost $500 billion a year in lost economic growth and productivity, according to one study by economists Victoria Fan, Dean Jamison and Lawrence Summers.[11]

A country's disease profile can affect direct foreign investment and trade flows. Even a contained outbreak that doesn't directly threaten human health may deliver a long-tail financial hit. Outbreaks of bovine spongiform encephalopathy, commonly known as "mad cow" disease, have been one of the worst public policy crises the British Government has faced during the postwar era.

The progressive neurological disorder of cattle that results from infection by a virus-like transmissible microorganism that transforms normal prion proteins into a disease-causing shape reached epidemic proportions in the mid-1990s. The crisis resulted in the slaughter of millions of cattle at home. The European Union slapped a ban on British beef exports that lasted 10 years, only ending in 2006.[12]

So I put this to Metabiota CEO Madhav: What would a really off-the-charts pandemic look like? Madhav calculates catastrophic risk for a living. She thinks about the unthinkable. A Yale University-educated epidemiologist and data scientist, Madhav and her team of emerging disease experts, computer programmers, actuarial experts and social scientists have created one of the world's most sophisticated pandemic models, shaped by AI disciplines like natural language processing and machine learning.

Drawing on a database that spans 2,800 outbreaks and some 150 pathogens over the last century, Metabiota's platform is designed to detect emerging diseases as they surface, even before public health authorities are aware of them.

Its surveillance system scrapes data from online media reports, travel data, company filings and government documents looking for information signals about emerging contagions in the voluminous noise of the global datasphere.

Have media reports spiked for certain symptoms like fever, nausea and chills and from where? What are the population and social mobility dynamics in that geographic region? How many airports are within a 100-mile range and where are those flights coming from and heading to?

The company also assesses how well individual countries are prepared to handle fast-moving health emergencies and the probable level of public panic during one. Pandemics cause massive social and economic disruptions. So Metabiota created a Pathogen Sentiment Index, a tool that predicts what companies and insurers can expect in the way of wealth-destroying, herd behaviors triggered by evolutionarily wired reactions to contagions.

Nathan Wolfe, Metabiota's founder and a former epidemiology professor at the University of California, Los Angeles, spent years living in West Africa trying to study viruses that jumped from wild animals to human populations. His 2011 book, *The Viral Storm*, in many ways anticipated the arrival of a virus like COVID-19.[13]

The company, which has a big on-the-ground research presence in Cameroon and the Democratic Republic of the Congo, played a big supporting role in the U.S. response to the 2015 Ebola outbreak. Yet when the crisis passed, the funding from Washington quickly dried up. "It turned out many federal agencies had a very short memory when it came to the potential posed by an epidemic," recalled former CEO Bill Rossi.[14]

Metabiota then pivoted to the insurance market, just as the industry viewed infectious disease as an escalating risk exposure in a world experiencing population growth, breakneck urbanization, shrinking wildlife habitats and growing microbial resistance to existing drugs and treatments.

The company started building out its modeling capabilities and teamed up with the German reinsurance powerhouse Munich Re Group and insurance broker Marsh to develop a product called PathogenRX, a new insurance solution offering to protect businesses from the disruption and financial hit of an epidemic.

When reports of a mysterious pneumonia circulated in China in the final weeks of 2019, Metabiota's system flashed red, as the company's AI-powered detection system assessed the probable severity of the outbreak, viral dynamics, fear levels and travel patterns.

Metabiota's pandemic prediction platform correctly forecast the high risk of the contagion in Wuhan spreading to Japan, Thailand and Hong Kong one week before confirmed cases were reported in those countries.

"We have the tools right now that could get us away from just reacting to pandemics," Madhav said, referring to the scanning power of AI technologies and data analytics. Beyond that, though, the scientific and government world's thinking about microbial threats needs to be far more expansive than just the biology involved. "Pandemics are a multi-faceted problem that require a multi-disciplinary response."

As destructive as the COVID-19 ordeal has been, it's on the lower end of the spectrum of truly deadly outbreaks we could conceivably face in the decades ahead. By the end of 2021, SARS-CoV-2 had killed more than five million, according to official tallies of confirmed deaths. However, the actual number may be double that. A study in May, 2021, placed global deaths at about seven million, or twice the then official total, based on an analysis of excess fatalities that appear to deviate from average mortality statistics.[15]

Even so, other potential viruses could do far worse damage, according to Metabiota's epidemiological models. Imagine a novel respiratory virus or influenza that's highly transmissible, but also more lethal and molecularly complex than SARS-CoV-2.

Vaccine development would be more challenging. Metabiota has run hundreds of simulations with that kind of pathogen in mind, with various parameters about its reproduction rate, case-fatality ratio and the length of time needed to develop an effective vaccine.

"The events we looked at are caused by pathogens that have respiratory spread, such as pandemic influenza and emergent coronaviruses," she explained. The average number of expected deaths was 264 million globally, ranging from 117 million to as high as 545 million, according to the 20 worst-case, multi-year pandemic scenarios simulated by Metabiota's models.

In the current crisis, we had effective vaccines being injected into the arms of citizens on an emergency authorization basis in about a year from the initial outbreak. We may not be so lucky next time. In 30 percent of the cases simulated by Metabiota, there wasn't a vaccine available, just as we don't yet in the real world have a vaccine for HIV/AIDs or Zika.

AI is also revolutionizing and speeding up drug development, which has taken on added urgency in this new age of pandemics. Vaccines are one of humanity's greatest scientific accomplishments. Few human endeavors have done more to combat infectious diseases and reduce global mortality rates.

Yet keeping the world safely immunized from an expanding list of fast-mutating pathogens is a struggle without end and requires a radical departure in vaccine development, a fascinating challenge that we'll turn to next.

Chapter 4

Vaccines Anytime, Anywhere

In 2015, a London-based artificial intelligence firm called DeepMind Technologies garnered international attention when its computer program bested one of the world's leading professional players of a thousands of years-old strategy board game called Go.

The rules of this ancient Chinese board game are simple, but the strategic choices involved are far more complex than chess. Starting with an empty board, players place game pieces called stones on open intersections of a grid. The stones don't move, and the aim of the game is to control more territory than your opponent by surrounding and capturing their positions.

In chess, there are perhaps 20 possible opening moves; with Go, there are 361 choices. The game requires visualizing a number of moves ahead, pattern recognition, and calibrating short-term tactics versus long-term strategy. Because of the bewildering number of decisions involved, the Chinese game had proved elusive for computer programmers to model. That is until DeepMind, now a subsidiary of Google parent company Alphabet Inc., developed a machine learning program called AlphaGo.[1]

"They built an algorithm using something called reinforcement learning that allowed them to play the game differently," former Google chief executive officer and executive chairman Eric Schmidt explained to me in an interview. "The interesting thing is that they played the game differently and came up with new moves that hadn't been discovered in thousands of years."

Now imagine that kind of analytical, predictive power trained on vaccine and drug development to fight infectious and antibiotic-resistant diseases.

DeepMind followed up AlphaGo with another AI program called AlphaFold, which has deciphered one of biology's most vexing challenges: predicting a protein's three-dimensional shape from its amino-acid sequence.

Google acquired DeepMind, co-founded by child chess prodigy and neuroscientist Demis Hassabis, in 2014. Hassabis had assembled a team with expertise in machine learning, advanced algorithms and neuroscience, and in those days Google was on the prowl for AI talent and startups worldwide.

Proteins are the building blocks of life, but are very complicated to model. What a protein does is determined by the unique amino acid sequence embodied in its three-dimensional folded structure, which affects its behavior and ability to bind to cells. Understand a protein's structure and you'll understand its function to a large extent.

Until recently, figuring out the exact geometry of a protein required painstaking lab work using techniques like X-ray crystallography and, in recent years, cryo-electron microscopy. By training an AI program to analyze proteins and predict their shape, AlphaFold has potentially streamlined this entire process and opened the doors to much faster vaccine and broader drug development.

In a protein modeling competition called the Critical Assessment of protein Structure Prediction (CASP), a team of DeepMind researchers turned heads in 2018 with the accuracy of their protein structure predictions. The company kept making refinements in the ensuing years, improving the program's analytical power.

In mid-2021, DeepMind and the European Molecular Biology Laboratory open-sourced a database of the predicted structures of the roughly 20,000 proteins expressed in the human genome and those of 20 model microbes, including *Escherichia coli* and yeast.[2]

Such a database is a remarkable achievement in the new era of AI-enabled biology, according to Schmidt, a founding partner in a tech and life sciences venture fund called Innovation Endeavors. "Proteins are essential to how life works, how life communicates," he said. "If you want to stop something from happening, or alter or make something stronger, this map will allow you to do it. Imagine the implications for drug discovery, for human health and longevity."

Knowing a protein's structure is crucial knowledge for structural biologists and virologists trying to develop vaccines and treatments to disable dangerous pathogenic viruses. AlphaFold's deep learning system can accurately predict a protein structure when no structures of similar proteins are known, a process known as free modeling.

After the DNA sequence of SARS-CoV-2, the virus that causes COVID-19, had been published online in early 2020, researchers worldwide focused on its crown-like protein shapes as a possible target for vaccines. Understanding how the virus' spike protein attached to healthy human cells and could be disabled turned out to be a valuable strategy in the successful development of vaccines.

Yet there's still much we don't know about the virus' molecular machinery and its ability to damage human health over the long-term. AlphaFold provided accurate predictions for lesser studied protein structures of SARS-CoV-2 that have been published in journals and shared with front-line vaccine researchers and clinicians.[3]

Building on these advances, Google parent Alphabet has launched a subsidiary called Isomorphic Labs that takes an AI-first approach to new drug discovery that will be run by DeepMind CEO Hassabis. "I think biology can be thought of as an information processing system, albeit an extraordinarily complex and dynamic one," Hassabis wrote in a blog post announcing the launch of Isomorphic. "Just as mathematics turned out to be the right description language for physics, biology may turn out to be the perfect type of regime for the application of AI."

AI-generated vaccines and treatments likely will play a vital role in humanity's future encounters with infectious diseases. The faster that we decipher a viral protein structure, the more expeditiously we can prototype drugs to stop it in its tracks. Luckily for us, researchers understood quite a bit about the spike protein structure of SARS-CoV-2, which emerged from a well-studied family of viruses called coronavirus.

As we confront future epidemics, deep-learning programs may be powerful tools in the broader field of "reverse vaccinology." By rapidly mapping out the proteins of an infectious virus or bacterium through computational analysis, we can more quickly design vaccines and antibiotics to disrupt their spread.

That kind of molecular intelligence of our microbial adversary is also very useful to a class of medicines called monoclonal antibodies derived from the cells of an infected host that are designed in labs and mass produced in factories.

Like the natural antibodies produced by our immune system, these designer molecules attach to a target protein of a virus and take it out of action or flag it for destruction by the immune system.

During the pandemic, companies such as Eli Lilly and Company, Regeneron Pharmaceuticals Inc., and GlaxoSmithKline Plc. and partner Vir Biotechnology Inc. developed monoclonal antibody therapies that were approved in the U.S. to treat mild to moderate cases of COVID-19.

Finding the molecular vulnerabilities of SARS-CoV-2 is also the aim of a class of drugs called nucleoside analogues that target the virus' reproductive mechanism. Merck & Co. has developed an oral antiviral medication called molnupiravir that introduces errors into the genetic code of the virus, preventing it from spreading in the body. Pfizer's antiviral pill, whose brand name is Paxlovid, is part of a pharmaceutical group called protease inhibitors that block an enzyme the coronavirus needs to replicate.

One of the most bedeviling facets of the COVID-19 crisis has been the emergence of wave after wave of mutated variants that threaten to make the first-generation of vaccines rolled out in late 2020 and early 2021 obsolete. Computational biologists are exploring ways for machine learning and DNA-sequencing technologies, in a process called deep mutational scanning, to help drug researchers better understand how viruses replicate and their potential to mutate.[4] That kind of precognition of not-yet evolved microbes could save lives.

Structural biologists believe the next challenge will be using AI to understand how several viral proteins work together in complexes to bind to healthy human cells, replicate, and evade our body's immune system. We still have a long way to go, but our increasing awareness of how proteins work with technologies like DeepMind's AlphaFold could have big implications for vaccine development in the years ahead.

"Will AlphaFold 2 help us out in future pandemics? These impressive advances provide a major leap forward in the ability to determine structures of single proteins," Oxford University biochemistry professor Matthew K. Higgins wrote in an essay on the DeepMind feat. "However, to make major contributions to structure-guided vaccine immunogen design, more is needed, including prediction of structures. It would be unwise to bet against the minds at Deepmind to make these advances in the future."[5]

Radical Step Change

The COVID-19 calamity is the greatest public health disaster in more than a century and the sharpest financial shock to the global economy since the Great Depression. Yet its destructive power overshadowed a remarkable scientific achievement: The rollout of effective and regulatory-approved vaccines within a year of genetically sequencing the SARS-CoV-2 virus.

In a field where the pace of new drug development is measured in decades, that's a radical step change in vaccine development cycles. New innovations such as messenger RNA (mRNA) and viral vector technologies delivered on their promise precisely when the world needed them.

A partnership between Pfizer and a German biopharmaceutical company called BioNTech produced a highly effective mRNA vaccine that was approved for emergency use in about eleven months. In roughly the same time frame came vaccines from the American biotech firm Moderna and a tie-up between AstraZeneca and Oxford University. Other vaccines, if somewhat less effective, were developed by Johnson & Johnson, Russia's Gamaleya Research Institute of Epidemiology and Microbiology, and Chinese players such as Sinovac Biotech, China National Pharmaceutical Group Corporation, commonly known as Sinopharm, and CanSino Biologics.

In the history of vaccine development, that kind of rapid turnaround is unheard of. And yet, it still may not be fast enough. We will need even more accelerated timelines to keep up with microbial threats in the decades ahead, according to Richard Hatchett, a physician and Chief Executive Officer of the Coalition for Epidemic Preparedness Innovations (CEPI). During the time it took Pfizer and BioNTech to produce their vaccines, there were 1.5 million officially confirmed deaths due to COVID, according to Hatchett.

CEPI aims to compress vaccine development timelines to 100 days. Getting there will mean reimagining how we make, manufacture, distribute and deliver vaccines. One crucial aspect to pulling that off will be having the same kind of research foundation for other infectious disease threats that we had in place with coronaviruses thanks to vaccine work on earlier outbreaks such as SARS and MERS.

With the body of research already in place with COVID-19, "we got off the gate like (Jamaican Olympic sprinter) Usain Bolt in terms of responding to coronavirus," Hatchett said in an interview with a Bloomberg colleague. "We're not ready to do that with other viral families," he said, "That presents concerns."

His group plans to raise $3.5 billion to fund a five-year plan to invest in next-generation vaccine approaches, establish worldwide manufacturing and supply networks and work with regulators to streamline and accelerate vaccine testing and drug approval processes.

CEPI was inspired, in part, by the fact that it took 10 years to develop an effective Ebola vaccine while thousands died in West Africa. Launched in 2016 with initial financial backing from the Bill & Melinda Gates Foundation, the Wellcome Trust in the United Kingdom, as well as the governments of Norway, Japan and Germany, the group is considered the one of the most ambitious vaccine development initiatives ever.

Hatchett has held senior White House advisory roles in the administrations of George W. Bush and Barack Obama. He also served as the chief medical officer with the U.S. Biomedical Advanced Research and Development Authority, overseeing countermeasure programs against chemical, biological and pandemic threats and leading the development of vaccines, therapeutics and diagnostics for emerging diseases.

Now based in Oslo, Hatchett travels around the world meeting with policymakers, scientists and pharmaceutical executives driving home the urgency and enormity of the task ahead. Vaccines are among the greatest public health achievements in modern times. They have saved, and continue to save, millions of lives.

Yet in an intricately networked world, in which pathogenic microbes jump species and spread exponentially, our vaccine arsenal needs to be flexible, fast and globally accessible. We can no longer passively react to new pathogens, Hatchett and others argue, then develop a vaccine to respond.

Instead, Hatchett sees the emergence of vaccine-making platforms that can quickly develop agile responses to pathogens across the 25 viral families and approximately 225 diseases that are known to infect humans and cause diseases. He thinks we need to spend the next 5 to 10 years identifying other potential spillover viral threats still lurking in nature.

The idea is to develop a pipeline of prototype vaccines that, with some quick modifications, can be quickly rolled out. This new approach to vaccine creation also needs to be synchronized with AI-powered pandemic surveillance systems, manufacturing and supply chains, as well as healthcare systems worldwide.

Hatchett is convinced there will be more pandemics in our futures in the decades ahead. "Stuff is going to come across the animal-human interface," he explained. "There is ahead of us in this century a pandemic that would make COVID look like a mild pandemic relative to what is possible."

In Hatchett's view, we are going to have to invest in biosurveillance capabilities and technologies to produce new vaccines and therapeutics much more rapidly than ever before. "If we stay focused," he explained, "we can dramatically reduce or even eliminate the risk of future pandemics if we want to. I think we have to, because they're an existential threat to modern society."

Mists of Time

The history of vaccine development extends back through the mists of time and into the ancient world. Before the World Health Organization declared smallpox eradicated in 1980, that disfiguring and lethal disease caused by the variola virus had tormented humanity for millennia, killing about one-third of those it infected and an estimated 300 million in the twentieth century.[6]

The exact viral origin story of smallpox is lost in the prehistory of human civilization. Yet there's some evidence that the ancient Egyptian world coped with the disease and the mummified head of the pharaoh Ramses V, who is believed to have died around 1145 BC, shows evidence of the scourge.[7]

The malady is described in ancient Sanskrit texts in India; the Liao Dynasty in tenth century China appears to have tried immunization strategies, such as rubbing a powder made from small scabs into the scratches on the skin's surface.

Centuries later, British physician and scientist Edward Jenner realized that people infected with the cowpox virus did not catch the much more deadly smallpox disease. In 1796, he inoculated a young boy with puss from cowpox lesions. Jenner is credited with discovering a potential vaccine strategy for smallpox. Up to that point, the only means to combat the scourge was a risky technique called variolation, or intentionally infecting healthy people with viral matter from smallpox patients. That was a bit like playing viral Russian roulette, though, as some exposed patients became seriously ill and died.[8]

In the decades that followed, the basic approach of using a weakened or inactivated microbe against itself didn't change all that much. Researchers typically had to isolate a pathogen, figure out how to grow it in chicken eggs, purify it, activate it and then formulate a safe version of it.

It was a laborious, career-long endeavor in some instances. A vaccine program could easily take more than a decade if you factored in the time needed for initial research, formulation, pre-clinical and clinical testing, and regulatory approval.

It took more than two decades of research by American medical researcher Dr. Jonas Salk and others before a safe and effective vaccine had been developed in the mid-1950s against poliomyelitis, the crippling neurological disease commonly known as polio.[9] In the early 2020s, the malady has been virtually eradicated in most parts of the world, but it is still endemic in Afghanistan and Pakistan.

Today, drug developers can design a prototype vaccine on a desktop computer in a matter of hours. What's changed is our understanding of genetics, our ability to reengineer the molecular mechanisms that run our cells, and the arrival of more computing and sequencing technologies that are becoming more powerful and more affordable every year.

That said, it took an enormous amount of money to jump-start vaccine development in the latest crisis. By far the biggest dollop of cash was the $18 billion the Trump Administration allocated for Operation Warp Speed, a crash development program initially focused on various vaccine programs for late-stage clinical research, manufacturing and purchasing agreements.[10]

The U.S. required that vaccine-makers secure supplies of any successful drug for the American population. Even so, the Trump Administration deserves credit for the broad, catalytic impact that the Warp Speed program had on the global response to the COVID crisis. With that kind of muscular emergency funding and regulatory support, drug companies could run multiple drug trials in parallel and on an accelerated basis.

Yet perhaps even more fortuitously, drug companies were able to leverage years of previous research on other coronaviruses and mRNA technology that was just coming to fruition. It takes nothing away from this triumph to point out that there was an element of sheer luck at work as well.

Traditional vaccines have worked with a straightforward approach: inject a weakened or dead form of a pathogen into a patient to stimulate his or her immune system. Toxins and foreign substances have molecules called antigens that when recognized can induce the immune system to attack with antibodies and other measures.

The vaccines developed by Pfizer and BioNTech, as well as Moderna, are genetic vaccines that take a different tact. They introduce an mRNA sequence into the body that instructs a vaccinated person's own cells to make a harmless version of a protein from a pathogenic microbe that triggers an immune response.

Medium Is the Message

The genesis of mRNA research winds back to 1961, when researchers at the Institut Pasteur in Paris discovered a ribonucleic acid (RNA) that copied genetic instructions from DNA to make the proteins essential to human biological life, governing everything from breathing and digesting to thinking.[11] It was a messenger RNA in the sense that it gave human cells the exact recipe to synthesize the proteins.

This critical mRNA molecule wouldn't be engineered in a lab until the mid-1980s. Nevertheless, researchers became intrigued by the idea that a synthetically designed version of it could be injected into patients and transform their cells into an on-demand drug factory for all sorts of potential therapies.

One researcher who saw that promise was Katalin Karikó, a Hungarian-born biochemist, who joined the University of Pennsylvania's School of Medicine in 1989 to research mRNA therapies. She saw the technology as a way to treat cystic fibrosis and strokes, but her work failed to secure sustained funding early on. The other issue was that mRNA was unstable to work with and sometimes triggered an inflammatory, even fatal, reaction when injected into test animals.

She eventually teamed up with Drew Weissman, an immunologist and HIV vaccine developer at the National Institutes of Health. The duo came up with a critical breakthrough in 2005.[12] By tweaking one of the mRNA's four building blocks called nucleotides that code for a protein, they could reduce the inflammation and increase the potency of the molecule.

They patented the technology in 2012, which was later licensed by BioNTech (where Karikó now works as a senior vice-president) and Moderna co-founded in 2010 by stem-cell biologist Derrick Rossi. For their COVID drugs, both companies designed an mRNA that instructs human cells to make harmless versions of a coronavirus' spike protein. That in turn triggers the immune system to make antibodies that will recognize and attack SARS-CoV-2.

One advantage of this technology is its flexibility. "We can target the mRNA to certain cell types and perform different kinds of treatment," said Karikó, who together with Weissman, granted interviews for a Bloomberg documentary about mRNA and cancer research.[13] BioNTech has an mRNA cancer vaccine in clinical trials and hopes to deliver its first treatments after securing approval by the middle of the 2020s. On the flip side, mRNA vaccines are very tricky to store and transport, given the very low temperatures required to keep these drugs stable.

The next big act for mRNA research may be a full-spectrum vaccine against the coronavirus viral family, responsible for so many epidemics this century, as well as dangerous flu strains. "I think the pan-influenza and pan-corona vaccines will be the most exciting mRNA vaccines to come out in the future," according to Weissman.

Aside from mRNA research, COVID-19 vaccine development also received a huge tailwind from preexisting work in both the U.S. and United Kingdom on the coronavirus that causes the Middle East respiratory syndrome (MERS), first reported in Saudi Arabia in 2012.

Researchers at the U.S. National Institute of Allergy and Infectious Diseases had explored ways to stabilize one of its spike proteins and trigger a more potent immune response. Oxford University scientists had also been working on a MERS vaccine.

Those earlier research initiatives proved valuable, particularly in the development of viral vector vaccines. In this approach, a modified version of a virus is the delivery-vehicle (or vector) of genetic instructions to our cells, with the same ultimate aim of firing up our immune system.

At Oxford, years of research by Sarah Gilbert and Adrian Hill at the Jenner Institute paid off with a clever approach for a COVID-19 vaccine. The vector in this case was an adenovirus that causes the common cold in chimpanzees. Oxford researchers had been using it to work on a MERS vaccine, but quickly shifted their attention to SARS-CoV-2 once it had been genetically sequenced.[14]

As they pivoted to COVID-19, they knew two things that gave them optimism they might succeed. First, the modified chimp adenovirus had proven safe in human tests with a MERS vaccine. Secondly, there was a fair amount of genetic similarity between the spike proteins of MERS and SARS-CoV-2.

The Oxford team modified the chimpanzee adenovirus with genetic material for the spikey protein on the surface of the virus that causes COVID-19. It was also altered so as not to replicate in humans. When injected, the adenovirus triggered antibodies and white blood cells called T-cells that proved effective against SARS-CoV-2 and would perform well enough in human trials to get emergency authorization approval in the U.K. at the very end of 2020.

Johnson & Johnson successfully adapted the same approach but used a human adenovirus as its message-delivering, molecular emissary to human cells. In some rare cases, these two adenovirus vaccines caused blood clotting, but the side effect risk was deemed statistically acceptable by most regulatory authorities.

Pathogens Incognito

Beyond COVID, the bigger, looming question is whether we have the technologies in place to deal with a completely brand new pathogen from scratch. We are not there yet, but might be closer by the end of the 2020s to having the ability to improvise and create vaccines on demand to deal with future threats from completely novel pathogens that nature might throw at us.

The phrase "Disease X" brings to mind some sort of deadly extraterrestrial microorganism or a bioweapons experiment gone terribly bad. Yet in 2018, that mysterious name was added to the World Health Organization's annual list of killer pathogens alongside Nipha, Ebola, Zika and SARS of urgent research priorities.

Somewhere out there, perhaps in the inner recesses of a bat cave in central China, among the shorebirds and gulls migrating from South America or the swine livestock population in the United States, a pathogenic virus, bacteria or fungus is silently circulating, waiting for a right molecular mutation and transmission opportunity to jump species and trigger the next pandemic.

With that unknown threat in mind, Imperial College London is leading a consortium to develop a platform called RapidVac that could produce flexible and synthetic RNA vaccines that would work against an array of viral diseases as diverse as H1NI influenza, rabies and the Marburg virus.[15]

Just like mRNA vaccines, these RNA-based drugs would deliver a message to human cells to make a particular protein based on the genetic sequencing of a target pathogenic virus that would activate the immune system and ward off infection.

The difference is that this next-generation approach would target more than one virus, while retaining a feature of RNA that allows it to produce multiple copies of itself via its so-called amplification machinery. If successful, these self-amplifying RNA (saRNA) molecules would require smaller doses while still being effective. A one-shot vaccine that worked across different viruses would allow manufacturers to scale up production and reach more arms in less time.

CEPI is investing in the RapidVac system, as well as another technique known as molecular clamping being developed by researchers at the University of Queensland.[16] This vaccine approach could also potentially work across many types of viral infections. It focuses on the surface proteins of a pathogenic virus that bind to and then hijack a human cell's replication machinery to make copies of itself and spread.

Vaccine developers are trying to zero in on these surface proteins. If you can synthesize a version of it that the immune system will recognize as a threat, you can develop a vaccine to attack it. The challenge is that these surface proteins are unstable and shape-shifting, often evading detection of disease-fighting immune cells. The Australian researchers have developed a way to "clamp" down on the original forms of these virtual proteins and make copies of them in a way that the immune system can readily recognize.

The research still has some ways to go, and wasn't developed enough to be helpful with COVID vaccines, but could be a game changer down the road. The approach would apply to any virus, whose proteins have been genetically sequenced. The University of Queensland team think their platform could someday produce a vaccine suitable for testing in four months from the identification of a pathogen.

At the same time, vaccine manufacturing technology is changing. The German biopharmaceutical company CureVac is developing portable RNA vaccine printing technology. The printer still in development can produce mRNA vaccine candidates for multiple pathogens and more than 100,000 doses in a few weeks. They could be deployed to infectious disease hotspots and parts of the world that don't have vaccine-making infrastructure to save time and reduce costs.

Anti-Vax Diehards

Almost as challenging as making these next-generation of vaccines will be convincing individuals to actually take them. The anti-vaccine movement has been around pretty much since the arrival of vaccines.

The rollout of mandatory smallpox vaccines in late-nineteenth century England touched off mass protests and objections on religious, scientific and political grounds. The level of distrust in the medical establishment and the government in Victorian England among underprivileged segments of society also played a role.

Anti-vaccine sentiment in the age of COVID is also multi-faceted, ranging from concerns about the impact of genetically engineered vaccines rippling through our bodies to paranoid conspiracy theories.

A video titled *Plandemic* that blamed the COVID-19 contagion on pharmaceutical companies, Microsoft co-founder and philanthropist Bill Gates and the World Health Organization attracted eight million viewers on Facebook, YouTube and Vimeo. [17] Yet it's important to remember that the anti-vaccine movement is a many-splendored thing.

Some of us just don't like needles or the idea of foreign substances being injected into our bodies. Taking syringes out of the equation will help, some argue. Vaccines that can be delivered orally such as the ones for polio are easier to administer and transport and eliminate needle injuries to healthcare workers. Yet not every vaccine can be absorbed efficiently through the digestive tract.

Another approach are micro patches with tiny needles that inject the vaccine. Scientists at the Delhi-based Hilleman Laboratories in India have developed patches for routine immunizations like hepatitis B. The drugs are easy to store and transport, and can more easily reach rural, low-income families in India. Public pressure, meanwhile, on social media sites such as Facebook, Twitter and Google's YouTube have forced these companies to remove disinformation about vaccines and the nature of viruses.

The COVID-19 vaccine rollout is the largest and most complex ever attempted, and one full of logistical snafus and legitimate criticisms about the uneven distribution of immunizations worldwide. Rich countries like the U.S. and the United Kingdom had enough vaccines to cover their populations several times over. India and nations in Africa struggled to get their hands on these drugs as the pandemic spread through their populations.

A program called Covax aimed to secure equitable vaccine deployment worldwide. The idea that a multibillion-dollar alliance of international health bodies and nonprofits would use their buying power to supply poor countries is a good one. But in the end, rich countries and major pharmaceutical companies didn't deliver to the extent hoped for, due to supply constraints and political pressure at home to keep their populations safe.

Before the next pandemic, we'll need to sort out technology-sharing and manufacturing licensing agreements that can be implemented in an emergency so vaccine manufacturers worldwide can scale up and produce drugs the world needs. Temporarily lifting intellectual property rights on a successful vaccine is a rather blunt approach, but shouldn't be taken off the table.

Just as critical will be having the necessary surge capacity in the vaccine manufacturing supply chains for the various critical raw materials, syringes, sterile filters, vials and stoppers needed to immunize the world in a hurry. That may require government financial backstops to help companies scale up quickly. Nations stockpile weapons. We need to do the same kind of forward planning when it comes to vaccine deployment.

The other challenge is keeping an accurate count of who has been vaccinated and who hasn't, especially in parts of the world where distrust of governments runs high and administrative capabilities are low. If a dangerous virus lives on in isolated pockets of the world, nobody is safe.

Changing that will require investments in smart technologies and biometrics. A major Ebola outbreak took place in the Democratic Republic of the Congo in 2019. In neighboring Rwanda, 200,000 people were enrolled in a two-dose program for a new Ebola vaccine to contain the viral crisis.

To keep track, healthcare workers were given handheld scanners with cameras connected to the Internet. The biometric device scanned the irises of patients to create a secure, personalized immunization record.[18] Authorities also sent out text messages reminding patients to get their second dose as part of a drive by public health experts to raise immunization levels.

Privacy advocates may cringe, but in countries without reliable identification systems, biometrics are more reliable data and can be designed in a way that keeps medical information confidential. Leveraging the ubiquity of mobile phones offers a smart strategy to track immunization, vaccine efficacy and side effects in a fast-moving crisis.

There will likely always be individuals willing to exploit infectious disease crises for their own anti-vaccination ideological, political or financial agendas. However, the data show that immunization rates have improved markedly over the past several decades.

In the 1990s, immunization coverage in less-developed countries hit a wall and low-income countries had to wait an average of seven years for access to newer vaccines versus the rich world. Children in impoverished parts of the world were very vulnerable to measles, polio and pneumonia.

Today, some 81 percent of children receive routine vaccines, versus 59 percent in 2000, according to Gavi, the Vaccine Alliance. The number of children being swept away by vaccine-preventable infectious diseases has dropped by 70 percent over the last two decades.

Vaccines are among the greatest public health achievements of modern times. Yet the microbial world will continue to throw novel pathogens at us. It will take a multi-pronged defense involving high-tech surveillance, AI technologies and next-generation vaccine platforms to improve the world's pandemic preparedness.

All of this will take sustained and super-collaborative, international coordination. There is one organization that can tie all these strands together into a cohesive whole, but it will have to change its entire operating model to get there. Right now, the World Health Organization is underfunded and spread too thin. What follows is how it might be reimagined as our super sentinel for a pandemic age.

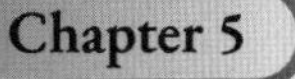

How the WHO Lost Its Way

Institutions like the United Nations, the International Monetary Fund and the World Trade Organization can be thought of as Rorschach tests for an increasingly polarized, angry, and globalized world. They reflect back the world view of their most avid supporters and acerbic critics.

To the idealists, they're our best shot at overriding our tribal, nationalistic impulses, and achieving the level of international cooperation needed to cope with this century's pressing challenges, from hunger and wealth inequality to ocean health and climate change.

The other camp views the global bureaucrats who run these organizations and their agendas with deep suspicion. They're empire builders, unaccountable power brokers. In the fever dreams of some conspiracy theorists, these institutions are controlled by unseen elites for nefarious purposes.

Then there's the World Health Organization, a multilateral body entrusted with promoting our collective health and well-being. The WHO is rightly credited with some incredible accomplishments such as helping vanquish smallpox, battling tuberculosis and polio, and drawing attention to the health needs of the undeveloped world. Research by its doctors and scientists over the decades have made us smarter about managing our public health.

Yet since the WHO's founding in 1948, the health policy landscape has changed significantly. In the 2020s, there are now dozens of entities promoting and funding better public health, from non-governmental organizations to multi-billion dollar philanthropic groups like the Bill & Melinda Gates Foundation, the Wellcome Trust in the United Kingdom and

Bloomberg Philanthropies. Their expertise is highly regarded, and they're far more specialized, well-funded and nimble than a lumbering, hydra-headed bureaucratic monster like the Geneva-based WHO.

With the increasing pace of infectious disease outbreaks this century, the WHO has been roundly criticized for mishandling the swine flu crisis in 2009, the 2014 Ebola flare-ups and a once-in-a-century pandemic like COVID-19 that broke out in China in late 2019.

Granted, the WHO is an ideal punching bag for national leaders seeking to divert domestic attention from their own policy failings or opportunistically pursuing their geopolitical agendas. We share the planet with trillions of microorganisms, some of which make us sick. That's just the human condition, or perhaps more accurately the microbial condition, given these tiny life forms essentially are the operating system of the biosphere.

Yet it's also true that the organization's own member states have widely divergent national interests, making quick and bold action sometime difficult. As we've seen, the WHO clearly was reluctant to confront China early in the pandemic over its refusal to share patient data and allow international experts into the country to investigate the origins of the SARS-CoV-2 virus.

Does the world still need the WHO? Of course it does. Few would argue that having an international institution focused on promoting global public health isn't a good thing. The issue is whether the WHO has the right structure, funding profile and influence to pursue that goal. If not, what do we want the WHO to do and how? What does success look like?

The WHO will never be able to keep us perfectly safe from infectious disease threats in the future. That's simply not realistic in an interconnected world with nearly eight billion people and the mind-boggling complexity of evolving microbial threats in the biosphere. It would be like expecting the U.N. to prevent human conflict and wars, or the IMF to offer failsafe protection against financial crises.

The WHO could, however, be more focused and serve as the indispensable nexus point in a newly fashioned information network designed to rapidly detect, assess and respond to future emerging infectious disease risks. Viral and bacterial flare-ups can be contained if confronted early and decisively enough. Yet it takes a comprehensive approach and the full backing of stakeholders throughout society.

COVID-19 raced around the globe because there wasn't anything approaching an effective global pandemic preparedness system in place, no early warning system, no mission control to coordinate actors worldwide in common purpose. Now that millions have died, some governments seem serious about investing in one and the WHO could play a critical role in helping shape and manage those efforts.

In September, 2021, the Biden Administration proposed a sweeping $65 billion, 10-year biosecurity plan to build up the nation's pandemic preparedness infrastructure to manufacture vaccines, treatments and tests more quickly. The White House compared the spending plan to the Apollo moonshot program of the 1960s.[1] If realized, it would be one of the biggest public health investments in U.S. history.

Other governments will need to make similar investments and they will all need to be coordinated to some degree, a role that an international entity dedicated to global health would seem well-suited to handle. The WHO is already leading efforts to create a new international treaty on pandemic preparedness and response.

The WHO and the Federal Republic of Germany have plans underway to establish a new global hub for pandemic and epidemic intelligence, data, surveillance and analytics innovation. The aim is to harness the potential of artificial intelligence and diverse data sources across multiple disciplines to move faster and more decisively in the face of emerging infectious disease threats.[2]

For all of its shortcomings, the WHO still has tremendous convening power during global health crises and brings quality research ideas and best practices to all corners of the world. Still, if the WHO is going to stay relevant and be more effective, even insiders concede the organization needs a rethink, if not an entire overhaul.

Since its mandate was broadened in the late-1970s beyond disease fighting to the promotion of universal health care for all, mission creep has made the global health organization arguably less effective. Its programs now span everything from preventing traffic accidents, obesity and domestic violence to promoting healthcare infrastructure and pharmacy systems.

Along the way, the WHO has waded into highly politicized policy waters, where it has less expertise or experience, and has often duplicated efforts by other organizations or national governments. The issue isn't whether these are important health issues that deserve consideration. They most certainly do. The question is whether there's a pressing need and comparative advantage for an international organization like the WHO to lead that debate.

A more focused mandate on big risks like communicable diseases and drug-resistant microbes plays to the WHO's core strengths. Unlike individual countries, the public health guardian is also an important advocate for low-income countries that don't have major pharmaceutical companies or ready access to vaccines and therapies in an emergency.

The failure to equitably share COVID-19 vaccine doses with the developing world is both morally abhorrent but also tactically inept with an evolving virus. The longer that big population centers in Africa, Asia and Latin America are unprotected, the more opportunities SARS-CoV-2 has to mutate into a variant that can outfox the major vaccines rolled out in 2020 and 2021.

Even bigger threats loom. If you ask epidemiologists and medical professionals what trend concerns them the most, one at the top of the list is the rise of drug-resistant microbes that make antibiotics ineffective and set us back decades in our ability to manage disease and infections. Humanity has made big strides in extending life spans over the past two centuries. Those gains are at risk if we enter a post-antibiotic age. It's worth quickly recalling what the world looked like before the international community cooperated on public health matters.

Brutish and Short

In 1800, Napoleon Bonaparte drove Austrian troops out of Northern Italy, Ludwig van Beethoven premiered his Symphony No. 1 in Vienna and slave-owner Thomas Jefferson won the presidency of the United States. The global population hovered around one billion. The average life expectancy worldwide? Somewhere between 35 years and 40 years.[3]

Extreme poverty, high infant and child mortality, as well as premature death among adults, were woven into the human condition and no part of the planet escaped that reality. Life spans hadn't improved much since

Roman times. By the middle of the nineteenth century, increased trade and travel with the East spawned cholera and other infectious disease epidemics across Europe. The first International Sanitary Conference was held in Paris in 1851.[4]

English physician John Snow, who investigated a cholera outbreak in the Soho district of London in 1854, traced infections to a contaminated water pump and well on Broad Street and suggested effective measures to stop its spread, one of the first instances of public health epidemiology.[5] That same year, Italian physician Filippo Pacini identified the *Vibrio cholerae* bacterium as the source of a cholera flare-up in Florence.[6]

Years later came the landmark work of German physician Robert Koch, considered one of the founders of microbiology. Koch discovered the anthrax disease cycle as well as the pathogen that triggers tuberculosis.[7]

Koch established a causal relationship between microbes and illnesses. His lab work showed that diseases came from microorganisms that could be isolated from a sick animal, cultivated in a lab and then transferred to a healthy one triggering the same symptoms and disease. As the nineteenth century drew to a close, the world's understanding about the nature of microbial pathogens had begun to shift.

Up to that point, most medical authorities subscribed to the miasma theory of illness that dated back to ancient Greek physician Hippocrates, who theorized that foul air carried disease to its victims.

Armed with this new knowledge, improved sanitation and nutrition, rising living standards and the antibiotic revolution, the twentieth century ushered in what epidemiologists now call the "great health transition" in the industrialized West. Recognizing the importance of internationally coordinated policy, the League of Nations formed in 1920 established an international public health arm.

The human carnage from two world wars notwithstanding, by 1950 average life expectancies had reached into the early 70s in the Nordic countries, the upper 60s in other parts of Western Europe and North America and the mid-50s in the Soviet Union.[8]

However, the great health transition never arrived in the big population centers of Asia or across Africa. That depressed the average lifespan worldwide to roughly 47 years in 1950. The yawning public health chasm between the rich and lower-income nations was breathtaking. In the middle of the last century, the average citizen in Norway could look forward to living 71 years; in Mali it was 26 and all of 36 in Africa more generally.

The World Health Organization came into existence in 1948, in part, to close that gap by combatting infectious disease. Spun out as a subsidiary of the United Nations, its central mission promoted the "attainment by all peoples of the highest possible health," as outlined in its founding Constitution.[9] It viewed health as a fundamental human right and a primary government responsibility. Communicable diseases were considered a common danger and threat to international security.

Early on, and with the backing of the world's major powers, the WHO set forth some ambitious aims, starting with the eradication of smallpox. In the mid-1960s, when that disease still killed two million annually, health authorities in the U.S. and Soviet Union cooperated on the distribution of vaccines to low-income countries.[10] That same decade the WHO backed mass public health campaigns against syphilis and leprosy, helped arrest a cholera pandemic in Asia and the Western Pacific, as well as a yellow fever outbreak in Africa.

In the latter part of the twentieth century, the WHO launched programs to vaccinate children worldwide against tetanus, measles, diphtheria, tuberculosis and called for the complete eradication of polio, a disease now very close to being vanquished. Other programs focused on HIV/AIDS.

As a result of the public health organization's work on containing communicable diseases, huge strides have been made since 1950 to lower infant and child mortality and improve human health. By 2020, average lifespans worldwide had increased by 55 percent to 73 years, with the biggest gains coming in the developing world. Life expectancy across Africa is now in the late 50s and early 60s. It's hard to imagine the world making those extraordinary gains without an internationally focused advocate like the WHO.

Third Rail

Public health is an emotionally charged and contentious political arena. It's expensive and governments need public support to underwrite their disease-specific programs and allocate healthcare. Nations differ markedly in their health system designs and their ability to pay for it.

Organized religions, still politically influential in many parts of the world, impose doctrinaire views about lifestyles, diets and sexual behavior on their followers. Disease outbreaks are socially divisive and can be existential threats to those in power.

Over the decades, the WHO has been caught up in turbulent geopolitical cross currents. The international organization's internal politics can be quite Machiavellian, too. It's a bureaucratic labyrinth, governed by an unwieldy World Health Assembly with delegates from more than 190 countries, an executive board of technical experts and a director-general, who's elected by member states and eligible to serve renewable five-year terms.

The WHO also has six semi-autonomous regional offices with their own bureaucracies and nearly 150 country offices. There's also a secretariat with a staff of doctors, technocrats, appointees and administrators responsible for carrying out programs. Taken together, the health group is a complex bureaucracy with turf battles and internecine rivalries just like any other sprawling international organization or multinational corporation.

During the 1960s and 1970s, decolonization swelled the size of the newly independent nations eligible to vote at the World Health Assembly. Universal health care and public health investments for the developing world then became far bigger priorities.[11]

In the early 1990s, after the breakup of the Soviet Union, came the backlash from wealthy, market-driven economies that controlled the WHO's budget and critically questioned the agency's evolution as a healthcare and social issues reform advocate. Some big donor nations wanted the WHO to refocus more on biomedical research and public health advice focused on specific diseases.

Rich-world member states were also far less willing to generously fund the WHO, and the organization started to rely more on private donors that tied their donations to specific programs. As a result, the health organization began losing its policy-making autonomy. Today, approximately 20 percent of its $5.8 billion, two-year budget is unconditional. The balance is project-tied donations by governments, foundations and non-state actors.

Over the last two decades, its primacy in global public health has eroded as well. The WHO still has a powerful voice in crises. Yet it's now but one actor among many in the international health ecosystem. The U.S. National Institutes of Health, the European Commission and United Kingdom Medical Research Council are big contributors to health research.

The Gates foundation has helped launch major vaccine distribution programs like Gavi, the Vaccine Alliance and the Coalition for Epidemic Preparedness Innovations (CEPI). Such philanthropies collaborate with the WHO, but have far closer ties to pharmaceutical multinationals and biotech firms, and have big ambitions of their own.

The WHO has been second-guessed repeatedly in the scientific and public health worlds for its glacial response to rapidly evolving outbreaks. In late 2013 and early 2014, Ebola infections emerged in a remote rural part of Guinea, which lacked infectious disease controls, and spread to nearby Liberia and Sierra Leone.

One of the first international organizations to investigate the West African hotspot wasn't the WHO, but Médecins Sans Frontières, or Doctors Without Borders, a Geneva-based group founded in 1971 that dispatches rapid-response medical teams to deal with disease outbreaks and other natural disasters. Despite the group's warnings about the severity of the contagion early on, national authorities in Guinea downplayed the threat, perhaps worried about the economic fallout from international travel bans.

The WHO in March, 2014, characterized the disease spread as "relatively small still," sent in a team but withdrew it several months later. The Ebola virus kept spreading and Médecins Sans Frontières characterized the situation as "out of control" and called for international intervention. The WHO didn't declare the outbreak "a public health crisis of international concern" until August of that year.

"WHO did not mobilise global assistance in countering the epidemic, despite ample evidence the outbreak had overwhelmed national and non-governmental capacities — failures in both technical judgment and political leadership," concluded a review by experts at the Harvard Global Health Institute and the London School of Hygiene & Tropical Medicine.[12]

The WHO has also suffered a reputational hit over its handling of the COVID-19 pandemic. However inept its own pandemic policy performance, the Trump Administration delivered a thunderbolt of a rebuke in May of 2020. Trump announced plans for the U.S., a major financial donor, to end its membership in the WHO, accusing the organization of cozying up to a China that failed to share data about transmissibility and properly investigate COVID-19's viral origins.

Trump's successor quickly reversed that decision, yet the Biden Administration has also questioned the WHO's unwillingness to pressure Beijing regarding the origins of a virus that killed millions. An independent panel report ordered up by WHO's director-general found fault with the health organization's slowness in responding to the pandemic, the biggest global health emergency ever on its watch, and called for major structural reforms.[13]

Critics in the public health world contend that the WHO has lost its way and needs to rethink its fundamental mission, funding strategy and internal structure. In a way, it has been disrupted by the velocity of outbreaks in an integrated world this century, the rise of better funded and more technologically sophisticated public health organizations, and a mission creep away from specialties where it offers a comparative advantage. The WHO isn't going to disappear and continues to do valuable work. The more interesting question is just how relevant it will be when it celebrates the centennial anniversary of its founding in 2048.

Split Personality

The cultural divide between the rich industrialized and low-income nation states has been with the international health guardian since the beginning. If there are two senior leaders emblematic of those contrasting outlooks, the WHO's current Director-General Tedros Adhanom Ghebreyesus and David Nabarro, a veteran British health diplomat and the organization's Special Envoy on COVID-19, come pretty close.

They've been rivals in the past. Tedros ran as a developing world outsider and beat out Nabarro and other candidates for the top job in 2017, with support from the African and developing world voting bloc at the World Health Assembly. He's the first African and non-physician to lead the organization.

Tedros clearly believes the WHO hasn't lived up to its original vision to deliver healthcare for all and that the international community has starved the WHO of the resources it needs to improve things. With most of its budget tied to projects dictated by its donors, the organization has ceded its freedom to act. When he ran for his position, Tedros openly questioned whether members really owned the World Health Organization any longer.

He claimed world leaders no longer abided by their obligations to report outbreaks to the WHO, fearing economically destructive travel bans by the rest of the international community. Tedros campaigned to restore more funding and power to the WHO. Yes, the health organization needed reform, but any big changes should be balanced against a competing goal of stability.

Nabarro had the backing of then-British Prime Minister Theresa May and the British medical establishment, a world leader in infectious disease research, genetics and biotech. Late in the 2017 campaign, things turned ugly when an ally of Nabarro's accused Tedros of underreporting cholera cases during three flare-ups when he was Ethiopian health minister. Nabarro said he never authorized the statement. Tedros called it a smear campaign.

Both agree the health institution needs to change, then again anyone with serious political ambitions inside the WHO would at this point. The organization's failures over the past two decades have inspired tomes of white papers and independent panel reports decrying the lack of independent funding at the agency, its slow response in crises and its overextended policy agenda, ranging from traffic safety to lifestyle maladies such as obesity and diabetes. To expect the WHO to cover such a wide agenda with sufficient depth and expertise is a stretch.

The global awakening about pandemic risk is an opportunity to refocus and empower the WHO. Yet it will take a coalition of both rich and developing nations to push through a credible reform agenda. Otherwise, this once-crucial institution would seem to face a future of growing irrelevance.

Tedros thinks the WHO needs more funding and authority and a recommitment to the world's less fortunate. Nabarro talks about the "power of partnering," and asserts that the WHO needs to regain its credibility as a smart investment on the world stage. In Nabarro's view, the WHO must first become a "supple, strong, confidence-inspiring organization that fits at the highest levels of the political process, then the money will come and we won't have to worry about budgets."[14]

The two hail from very different worlds. Tedros was born in Eritrea, then a province of Ethiopia and now a separate country, and earned advanced degrees in infectious disease and public health at the University of London and University of Nottingham. As head of a regional health bureau in his home province of Tigray, Tedros pushed for expansion of staffing and vaccines and delivered reductions in the cases of HIV and infectious meningitis.

Tedros greatly expanded the country's medical care as Ethiopia's health minister, and later served as foreign minister in a ruling coalition government led by the Tigray People's Liberation Front (TPLF). That party was founded in the 1970s as a Marxist-inspired guerrilla group fighting for the rights of the northern Tigray region that borders Eritrea against Ethiopia's central government.

The TPLF and a new government led by Prime Minister Abiy Ahmed have clashed militarily in the early 2020s. In Tigray, thousands have been killed and more than one million families displaced. It's not clear how welcome Tedros will be back in Ethiopia when his run at the WHO ends. In November of 2020, a senior army official branded Tedros a "criminal," who's trying to help the separtist group. Tedros has denied that.[15]

Early on in the COVID-19 crisis, Tedros and the WHO's leadership received withering criticism for not leaning on China to be more forthcoming about the emerging crisis in Wuhan. It is required to do so under the legally-binding International Health Regulations (IHR), set up in the wake of the SARS outbreak in China.

In that crisis in 2002 and 2003, Beijing's obfuscations allowed SARS to spread quickly around the Asia-Pacific region and Canada. The episode cast an unflattering light on Beijing and Chinese health minister Zhang Wenkang lost his job for mishandling the crisis.

In early January of 2020, as a viral epidemic raged around Wuhan, Chinese authorities told Tedros and his scientists that a new virus, SARS-CoV-2, wasn't being widely spread person-to-person, an assertion the WHO appeared to take at face value and tweeted out to the rest of the world on January 14.[16] By the end of the month, WHO declared a public health emergency of international concern, but only after considerable internal debate by its executive board.

China also kept international observers at a distance. In late January of 2020, a WHO team made a brief field trip to Wuhan. However, it took more than a year of protracted negotiations for another mission to get on-the-ground access to explore the possible origin of SARS-CoV-2. The scientists who arrived to China in January of 2021 weren't given access to raw patient data and were greatly hamstrung as a result, the WHO later complained in public statements. As a result, the world still doesn't know how the viral spread began or whether it's still circulating in the wild or via an animal carrier.

Throughout the pandemic, it had become clear that the Chinese weren't living up to either the spirit or the letter of the IHR regulations. In January of 2020, Tedros praised China effusively after returning back from talks in Beijing with Chinese President Xi Jinping. "Its actions actually helped prevent the spread of the coronavirus to other countries," said Tedros. By the end of March in 2021, the WHO leader had publicly rebuked China for its stonewalling on the pandemic origin investigation.

Tedros was somewhat philosophical about the political heat he and the WHO endured in early 2020. "I have been through difficult conditions and situations in my life that have sometimes been a matter of survival," Tedros wrote in an email to a reporter. "To be effective in situations, I have learned to calm down and focus on doing the right thing and making the correct decision."[17]

China aside, Tedros has been quite critical about the pharmaceutical industry and what he claims is the tendency of the rich world to hoard life-saving medicines in times of crisis. As Ethiopian health minister, he was appalled that it took 10 years for low-income countries to gain affordable access to antiretroviral drugs to treat HIV/AIDs.

Catastrophic Moral Failure

By the end of 2021, the same dynamic was playing out with COVID-19 vaccines. The U.S., United Kingdom and Western Europe enjoyed relatively high immunization rates and were trying to move on from the pandemic. Much of Africa and parts of Asia will almost certainly be living with the destructive impact of the virus for much longer. In early 2021, Tedros warned the world was on the brink of a "catastrophic moral failure — and the price of this failure will be paid in lives and livelihoods in the world's poorest countries."

Nabarro, a refined and experienced globe-trotting international public health diplomat for four decades, has held powerful posts at both the WHO and United Nations. In 2017, as he campaigned to run for director-general at WHO, Nabarro called for radical change at the organization in the aftermath of its Ebola failure in West Africa.

"WHO must change the way it manages health emergencies, as a matter of urgency. I have seen first-hand the scale of suffering if the response is not effective," he wrote in an opinion piece for the *British Medical Journal*. "It is clear that the WHO is still not ready for another Ebola, and that must change."[18]

Nabarro is very much part of the British establishment, having studied at the University of Oxford and the University of London and qualified as a physician in 1973. His father was the prominent endocrinologist Sir John David Nunes Nabarro. Another relative, Sir Gerald David Nunes Nabarro, served as a Conservative Party member of Parliament during the 1950s and early 1960s.

Early in his career, David Nabarro did various stints in community development and humanitarian relief programs in Iraq, South Asia and East Africa. As a senior health diplomat, he held posts focused on malaria, avian influenza, Ebola and food security.

In 2003, Nabarro nearly lost his life when a massive truck bomb detonated in front of the Canal Hotel in Baghdad, in a terrorist attack that killed 22, including the UN's Special Representative in Iraq, Sergio Vieira de Mello. "I had glass stuck in the back of my neck," Nabarro said at the time. "Then we started to hear the screams and moans. And that just went on and on."[19]

When I spoke to Nabarro in late 2020, he agreed there were things the WHO might have done better in the first weeks of the response to COVID-19. However, he described it as an "extraordinary achievement" for the WHO to have taken a mere month from the time the virus was sequenced to declare a public emergency by the end of January.

Nabarro thinks there's a fundamental misunderstanding internationally about the powers that WHO actually has in crises. "We're not able just to call the shots all the time," Nabarro explained. "We have to operate in a political context."

Public health officials think in terms of scientific causality and offer specific advice about what needs to be done to stop an infectious disease in its tracks. "Whereas a politician says, well, it's not quite like that. I've got to deal with parliamentarians here," according to Nabarro. "I've got to deal with businesses there. I've got to deal with angry communities there. I'm having to balance all the time."

Nabarro thinks we will be living with SARS-CoV-2 well into the 2020s. But it could open our eyes to new ways of doing things both at the WHO and in the world more broadly. "We're being given a chance as humanity to learn an awful lot very quickly. And I suppose the question is, what will we do with this new learning." Nabarro said. "Will we be better? Will we be stronger? Will we be fairer?"

WHO Reimagined

If the WHO is going to be central to the global health system in the years ahead, it will have to find a more defined role that really adds value in addressing the microbial infectious disease challenges.

First, it needs to embrace some reforms that have been obvious for decades and outlined in policy papers ad nauseum. The WHO has evolved into a patronage machine to some degree and there needs to be a talent upgrade away from political appointees to more technically trained experts across multiple disciplines at its country and regional offices. Some reformers argue the director-general should serve just one seven-year term, so her or his decisions are free of political shading.

The WHO should consider narrowing its focus to primarily infectious disease work, particularly providing strategic advice about how to lower the risk of outbreaks and manage health emergencies. After the WHO gets its own house in order, the organization might have a stronger argument for tighter control over its budget, more latitude to declare public health emergencies and unfettered access to disease hotspots.

More ambitiously, the WHO could reposition itself as the north star in a new constellation of international public health actors focused on preventing future pandemics. If the world truly wants to avoid future Ebola, Zika and COVID-19 shocks, it can no longer passively wait for a crisis to strike, then formulate a response.

Instead, we'll need an integrated and always-on system focused on four primary tasks: building microbial surveillance networks; enhancing programs to track and manage pathogen spillover risk from livestock and wildlife; designing more surge capacity into our healthcare systems and medical supply chains; and accelerating research and development into vaccine platforms, therapeutics and antibiotics.

As we've explored in earlier chapters, the building blocks are starting to emerge, thanks to breakthroughs in genetic sequencing and bioengineering. MetaSub and the Global Virome Project, among others, could allow us to develop an atlas of the most dangerous viral and bacterial families.

Pandemic intelligence platforms similar to those built out by AI data firms like BlueDot and Metabiota are early warning systems for new disease outbreaks, often picking up signs of trouble before governments and the WHO.

With the right array of Big Data platforms and standardized information networks, genetic sequences of dangerous microbes could be fed quickly to developers with access to vaccine platforms in a matter of days. Prototype vaccines ready for clinical testing and regulatory approval might someday be ready in a matter of weeks.

Imagine a world in which mobile vaccine printers and rapid diagnostic tests are air shipped to disease outbreak locations early in an infectious disease crisis. If needed, global vaccine manufacturing networks and supply chains, with built-in surge capacity, could be activated as well, giving us a fighting chance to make sure localized epidemics don't turn into devastating global pandemics.

It would take years of investment and plenty of sustained political will to make this a reality, yet most of the crucial technologies are within our grasp.

What's needed for this kind of massive undertaking is a central intelligence network with the kind of expertise and global perspective to make sure the pillars of this pandemic surveillance and rapid response system are working in unison and meeting the needs of the entire world.

Given the millions of lives lost to COVID-19, the public health benefit of robust pandemic preparedness is obvious. Yet in a world of finite resources and stretched government budgets, there's also a very strong economic rationale as well for policymakers concerned about a return on their investment.

The international consultancy McKinsey & Company ran some numbers on what a decade-long investment to improve the world's health security might look like. They came up with a price tag of roughly $357 billion over 10 years.[20] That's a lot of money, but also a fraction of the more than $20 trillion hit that the COVID-19 pandemic is expected to cost the global economy in lost growth from 2020 to 2025.

The frequency and lethality of infectious disease outbreaks have risen this century, due to rising populations, rapid urbanization and climate change. It's one of our biggest and most immediate twenty-first-century challenges and the poorest sections of the planet bear the greatest risks.

Having a robust system in place to mitigate the damage would seem in sync with the Tedros viewpoint inside the WHO, while the return-on-investment calculations should appeal to the more results-oriented types like Nabarro.

Seven decades ago, the WHO stepped in and helped raise life spans around the world. Today, one of the biggest threats to public health is pathogenic microbes that we don't see coming. A more focused and agile WHO would be an invaluable strategic coordinator against such risks.

Big scientific ventures often yield unexpected commercial spinoffs that make our lives healthier and more prosperous. Thus far, we've focused on the threats. Now, it's time to explore the opportunities that microbes offer to extend our lifespans with new drug discovery and perhaps even change the contours of modern capitalism.

THE OPPORTUNITIES

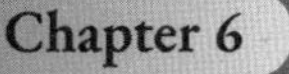
Chapter 6

Our Microbes, Ourselves

In May, 2021, a French biopharma company announced a remarkable research finding. A blind patient in his late fifties had his vision partially restored using a treatment developed by GenSight Biologics in Paris.[1] It's based on a technology called optogenetics, which uses light-sensitive proteins from microbes to influence the behavior of human nerve cells in the brain.

Optogenetics, pioneered in the mid-2000s by Stanford University neuroscientists Karl Deisseroth and Edward Boyden, currently a professor at the Massachusetts Institute of Technology, is now used by researchers in labs the world over.[2] It has fundamentally changed how scientists map the functions of the brain and explore therapies for Alzheimer's and Parkinson's diseases, as well as other conditions such as blindness.

In optogenetic studies, scientists genetically reengineer neurons so they express microbial proteins called opsins, which respond to light waves and are common in some types of algae and other microorganisms. One favorite lab opsin is found in *Chlamydomonas reinhardtii*, a single-cell alga that is activated by light. Other opsins shut down nerve cells when exposed to photons.

By selectively stimulating neural networks, neuroscientists gain a more nuanced view of how nerve cell branches in the brain communicate and interact. That's valuable to know, since some neurological diseases disrupt the flow of these electrical and chemical signals.

The field has leveraged decades of research into how microorganisms adapt to the cycles of sunlight and darkness on Earth. Billions of years of evolution have shaped how bacteria and algae react to solar energy and how microbes, more broadly, engage in photosynthesis on their own or symbiotically with plant life.

GenSight, co-founded by engineer Bernard Gilly and ophthalmologist Jose-Alain Sahel, has developed optogenetic therapies to treat hereditary vision disorders such as retinitis pigmentosa, which results in the loss of light-sensitive cells in the retina.

A human visual system is a sophisticated biological computer. Our retinas take in light via millions of photoreceptor cells and convert that energy into electrical signals that travel, by way of ganglion cells and the optic nerve, to a region of the brain called the visual cortex. There, signals are then processed into images. With certain retinal diseases, photoreceptor cells degrade or are otherwise disabled, resulting in vision loss.

GenSight researchers developed a gene therapy in which microbial proteins are injected directly into the eye's ganglion cells, bypassing the damaged photoreceptor cells, to make them receptive to light. Patients then wear high-tech goggles that capture light from the outside world. An algorithm built into the device processes and amplifies those images and then emits light in the interior of the goggles into the user's eye to stimulate the light-sensitive proteins that were injected into retinas during treatment.

After months of gene therapy and training, GenSight's blind patient, who was diagnosed about three decades ago with retinitis pigmentosa, started to see light patterns that seemed to vibrate. Eventually, he could locate objects placed on a table in front of him and identify a crosswalk on a street when he wore the goggles.

"For someone like me, who has mourned the loss of his sight, this is really extraordinary," the patient reported to a GenSight researcher. "It means that now, there's a lot of hope."

GenSight monitored the electrical activity in the patient's brain to verify that his visual gains weren't imaginary. "His visual cortex stimulation was tracked and it was consistent with patterns one sees with visual stimulation," according to company co-founder Sahel, who's also a professor at the University of Pittsburgh medical school. "That ruled out a chance, subjective experience."[3]

A New York-based company called Bionic Sight is also conducting human trials with a similar optogenetic approach. Other biotech firms such as Editas Medicine near Boston are using CRISPR gene editing, which exploits how bacteria counter invading viruses, to find effective genetic alterations to treat retinitis pigmentosa.

The Second Genome

Public awareness about the potential of microbial medicine awakened in 2012, when a global research consortium published the results of a five-year, $173 million study funded by the U.S. National Institutes of Health (NIH) called the Human Microbiome Project.

The aim: to genetically map the bacteria, viruses, fungi, and archaea that inhabit a typical body, the so-called human microbiome. Then-NIH Director Francis S. Collins likened the effort to define the human microbial makeup for the first time to "15th century explorers describing the outline of a new continent."[4]

Until the big advances in genetic sequencing in the mid-2000s and progress in a field called metagenomics, scientists had a limited view of how communities of microorganisms interact within our bodies. Most of the bacteria, for instance, that live in our gut are very difficult to culture and study in a lab. It's also a massive landscape to traverse. A 2019 study found an average of 46 million bacterial genes in oral and gut samples taken from 3,500 people. For researchers, that poses data and computational challenges.[5]

With shotgun metagenomic sequencing, researchers can now look for signals in the microbial noise. Virtually every cell in our body has DNA, which stores biological information in a quartet of building-block chemical bases: adenine, cytosine, guanine and thymine that are short-handed in science writing by their first initials A, C, G and T. The order or sequence of these bases, called nucleotides, are the biochemical source code for building and maintaining life.

In shotgun sequencing, the multitude of DNA found in a microbiome sample from our mouths or guts are isolated, extracted and then randomly broken up into sequencing fragments, not unlike scattered pieces of a jigsaw puzzle.

These individual pieces of genetic sequences are then analyzed by powerful computer programs looking for patterns, different combinations of

A, C, G and T, that are identical to each other. Overlapping sequence reads are combined, and the process repeats itself over and over again. Gradually, longer sequences, bigger sections of the genetic puzzle, fall into place and the computer pulls together a complete series of DNA sequences.

The data are cleaned up and refined to avoid duplication and then cross-referenced to genomic databases of previously identified microorganisms. Scientists sometime discover entirely new microbial species. While shotgun metagenomic sequencing has greatly expanded our vision into microbial ecosystems and is considered fairly accurate, it doesn't capture everything.

Some researchers prefer another sequencing strategy that focuses on a near-universal gene called 16S rRNA (as in ribosomal RNA) to guide their microbiome analysis. RNA has a core function in cells of encoding genetic information from DNA. It's a ubiquitous feature in all manner of microbes, serving as a useful barcode from which to identify individual microorganisms.

Another analogy might be a chronometer, just one that tracks evolutionary time and distance. By isolating the ribosomal RNA in a sample, studying its mutations and molecular structure and comparing it with other microbial genomic lineages, we can see where a species falls in the arc of biological life on Earth spanning billions of years.

Scientists have long known that microbes reside in our guts and help digest food. Yet it wasn't until the last two decades that we started to better grasp the complexity of the microbial swarms inside our body that extract energy from food, produce folic acid and vitamins like B2 and B12, manufacture neurochemicals that affect our moods, and fine-tune our immune systems.

The NIH microbiome study, as well as the crowd-sourced projects under the Microsetta Initiative that's collecting and sequencing microbe samples from volunteers in gut research in the United States and United Kingdom, has opened our eyes to the extraordinary community of tiny living entities shaping our individual destinies.

The American Gut and British Gut Projects (part of the Microsetta Initiative) are working with citizen scientists collecting samples worldwide to discern what lifestyle and health factors are associated with microbiome differences among populations.

One of the leading figures in human microbiome research is Rob Knight, the founding Director of the Center for Microbiome Innovation and Professor of Pediatrics, Bioengineering, and Computer Science & Engineering at the University of California San Diego. The Knight Lab has developed software tools and laboratory techniques for high-throughput analysis of microbiota. His team's research has made connections between microbes and obesity and inflammatory bowel disease.

The microbial universe living on our skin, in our gut, lungs and nether regions is a vast one. The human body is home to trillions of microbial cells.[6] Because of their infinitesimal size, they only make up 1 percent to 3 percent of the typical human body's biomass.

What they lack in heft, they more than make up for in impact. We carry around pathogens that under the right conditions can kill us. It's common to find bacterial strains of *Escherichia Coli (E. Coli)* and *Enterococci* in perfectly healthy people and it's unclear what chain of events will turn them into pathogenic adversaries.

Much of the microbiome we've coevolved with over the millennia is symbiotic or transactional. Microbes help us absorb nutrients and kill off disease-causing bacteria and viruses. In return, we provide an energy source and shelter.

Our relationship with microbes may begin as early as our fetal development and the type of birth one has dramatically shapes the microbial profile early in life. Babies who arrive via the birth canal and are breastfed have microbiomes that resemble their mothers; infants delivered via a caesarean (C-section) procedure tend to pick up the microbial profiles of their hospital and, some studies suggest, are at higher risk for asthma and obesity later in life.[7]

Lifestyles matter, too. The gut microbiomes of city dwellers in the industrialized West are less diverse than those of people living in more rural, low-income countries.[8] Children who grow up with pets, particularly dogs, seem to have fewer allergies and respiratory viruses.[9] As we age, our microbial ecosystem becomes less diverse and that correlates with greater frailty and inflammation.[10]

Microbes have been described as the second human genome, in the sense that the human microbiome is a supersystem that plays a critical role in human health. Optimists believe that unlocking its secrets could accelerate drug development and therapies that hold the potential to extend our lifespans.

During the 2010s, there was considerable excitement, and more than a little hype, about the dawn of a new era of personalized medicine that would cater to one's specific microbial and genetic profiles. Food companies promoted the prebiotic and probiotic dietary features of their products, which claimed to stimulate the growth of gut microbiota. A genre of self-help and diet books celebrating gut flora ensued.

Yet specialized dietary advice and supplements may not be all that effective given how individualized our microbiomes are in reality. While humans are typically 99.9 percent identical when it comes to our genetic makeup, our microbiota profiles can be 80 percent to 90 percent different from each other.[11]

The scientific consensus is that a widely diverse menagerie of gut microbes is better than a less diverse one. If so, that may be accomplished in most cases with the age-old advice to consume a lot of vegetables, fruit, and fiber.

Microbial imbalances, known as dysbiosis, have been correlated to diseases as varied as eczema, asthma, diabetes, obesity, inflammatory bowel disease, cancer, anxiety, and depression.[12] However, we still don't fully understand the structure, functions, and interactions of the networked community of microbes inside us.

The field still has years of discovery science ahead of it. Among the big questions to be sorted out: How do specific microbiomes take shape in individuals and why are they so diverse person to person? Why do some pathogens suddenly turn on us and under what specific conditions? What are the exact causal links between the microbial ecosystems and diseases like dementia, cancer, and antibiotic-resistant infections?

Gut-Brain Nexus

One area of microbiome science that's been illuminating is research into gut-to-brain communication networks. The human microbiome does way more than just break down food in our gastrointestinal (GI) tract that includes our mouth, pharynx, esophagus, stomach, and intestines.

Its impact is far-reaching, shape-shifting and influences multiple organs, including our brains. While our human genome doesn't change dramatically during our lifetimes, our microbiomes take on vastly different configurations as we age with potentially big ramifications for our well-being.

Experiments with germ-free mice have established a persuasive body of evidence about the intricate interplay among the microbiota, our brains, and behavior.[13] When the gut flora of patients with depression was transferred to germ-free animals, they displayed depressive-like behavior. Introduce microbiota from patients suffering Parkinson's disease or multiple sclerosis into these mice, and they tended to develop motor problems, as well as inflammation and autoimmune issues.

The communication networks between our microbiota and brains are multidirectional. Fear or social anxiety can cause the sensation of butterflies in our stomach. Alternatively, the microbial community inside our bodies sends biochemical messages that influence the central nervous system and brain. Microbes interact with our endocrine glands that produce hormones and regulate metabolism, as well as our immune system.

Yet some scientists think the real information autobahn of microbial, gut-to-brain communication may be the vagus nerve, the longest and most complex of cranial nerve networks that runs from the brain down the face and thorax, extending through much of the GI tract and down into the abdomen.[14]

The vagus nerve's sensory network and branches track chemical changes in various organs and transmit the information to the brain. Microbiota may communicate with the brain by stimulating the body's release of nutrients, hormones and peptides, which are small strings of amino acids and the building blocks of proteins.

The vagus nerve is also crucial to our sense of well-being, as it routes signals around the body to the brainstem, and from there into parts of the forebrain that govern emotion and motivation such as the amygdala, hippocampus, hypothalamus and substantia nigra. Animal studies over the past two decades have shown that if this vagus nerve is cut or disabled, test subjects suffer a range of neuropsychiatric problems.

Microbial communities are also complex drug factories churning out antibiotics and antidepressants. A growing body of scientific evidence suggests that gut microbes produce mood-altering neurochemicals like gamma-aminobutyric acid (GABA) that calms us, serotonin that elicits a sense of happiness and resilience, and dopamine that sharpens motivation, focus, and memory. They speak the same neurochemical language as our brains and the vagus nerve has receptors that pick up the chatter.

Studies have shown that supplementing the diets of lab mice with the bacterium *Lactobacillus helveticus* increased serotonin levels and reduced anxiety and depression.[15] Research is underway to better understand the links between human gut microbiota and infectious diseases, liver problems, gastrointestinal cancers, autoimmune syndromes, autism, and depression.

More broadly, there is evidence that microscopic life has been crucial to the evolution of the human brain. Some theorize that microbes have developed symbiotic strategies to effectively hack into our minds, shaping our neurochemical reward system to make us crave certain foods beneficial to their survival.[16]

Human microbiome science is full of interesting correlations about microbes and health, but not always the hard evidence explaining the causality behind them. That said, this field didn't really exist two decades ago and is now brimming with start-up companies boasting billion-dollar stock market valuations.

Investors are betting that microbial medicines are more than a passing fad. Rather, they may eventually prove to be essential tools in two big twenty-first-century health challenges: cancer and antibiotic-resistant microbes.

Tiny But Mighty

Cancer remains a leading cause of death worldwide, killing about 10 million people in 2020. The World Health Organization sees that figure climbing to 13 million annually by the end of this decade, as the overall global population continues to expand and age.[17]

Traditional cancer treatments such as surgery, radiology and chemotherapy continue to improve, but each have limitations. Now, there's research showing that bacterial cancer treatments may be useful in targeting tumors at the molecular level, enhancing the effectiveness of traditional cancer medicines, and rallying the immune system to recognize and destroy tumor cells. Let's be clear: The search for successful cancer treatments will almost certainly remain a challenging one in the decades ahead. That said, therapeutic bacteria is getting a serious look by cancer specialists.

Oncologists have long understood that bacteria can be an effective weapon against cancer. In the late-nineteenth century, William Coley, a young surgeon at New York Memorial Hospital, became intrigued by reports of cancer patients experiencing full or partial remission after experiencing bacterial infections from their wounds and surgery.

To test out his theory, he intentionally infected his cancer patients with a bacteria called *Streptococcus pyogenes.*[18] In some cases the strategy worked and the cancer regressed, but the treatment held risks if the bacterial infection, itself, became a health threat. He then developed a vaccine using metabolically dead versions of *S. pyogenes* and other bacterium called *Serratia marcescens.*

Known as Coley's Toxins, the treatment showed promise against various cancers, including sarcomas, carcinomas, lymphomas and melanomas. Yet Coley's natural remedy started to fall out of favor in the twentieth century as surgical operating rooms became sterile and infections of any kind were considered dangerous. By the time Coley passed away in 1936, radiation therapy had become a standard cancer treatment, while chemotherapy seemed like a promising future one.

Nearly a century later, we've learned a lot more about the relationship between bacteria and cancer, not all of it uplifting. The bacterium *Helicobacter pylori*, believed to be present in half the human population worldwide, is considered a risk factor for gastric cancer.[19] Some 33 other cancers have been linked to specific configurations of microbial DNA found in blood and tissue, according to researchers at The Cancer Genome Atlas, a database co-managed by the National Cancer Institute.[20]

On a more optimistic note, emerging bacterial anti-cancer treatments offer some potential advantages over chemotherapy, which often fails to penetrate deeply into the tumor's molecular structure and indiscriminately kills healthy cells, causing side effects like nausea, hair loss and compromised immune systems.

In contrast, some strains of bacteria like *Clostridium, Listeria* and *Escherichia coli* are quite adept at proliferating deep inside tumor cells.[21]

Genetically engineered bacteria are potentially excellent delivery vehicles for anti-cancer agents that can kill tumor tissue or restrain their growth. Clinical trials also show promising results when bacteria and chemotherapy are combined.[22]

One of cancer's most pernicious features is its ability to camouflage itself from and evade the human immune system. Bacterial cancer treatments have proven adept at exposing tumors to the human immune system's killer T-cells.[23] In addition, the risks associated with bacterial cancer treatments can be managed with antibiotics, which weren't widely available in Coley's day.

Some microbial cancer drugs are starting to wend through human clinical trials. A British biotherapeutic company called 4D Pharma, a leading player in biological products that treat tumors and neurodegenerative conditions, has found a strain of a gut microbe called *Enterococcus gallinarum* that boosts the immune system when absorbed by a receptor on intestinal cells.[24]

One of the company's cancer treatments based on *E. gallinarum* and known as MRx0518 is being tested with an immunotherapy drug called KEYTRUDA made by Merck Inc. that has had promising preliminary test results. If the final phase of tests meets expectations, 4D Pharma may be on track to receive regulatory approval for its microbial cancer treatment by the middle of the 2020s.

Other companies such as Seres Therapeutics and Vedanta Biosciences, both based in Cambridge, Massachusetts, are blending mixed strains of bacterial species to boost immune responses in cancer patients. In preclinical studies, Vedanta's VE800 treatment has shown the ability to help T-cells infiltrate and attack tumor cells.[25]

Cancer research aside, microbial science may also play a critical role in another looming twenty-first-century health challenge. Antibiotic-resistant superbugs have been infecting and killing people across the world in greater numbers in recent decades. These hard-to-treat bacterial and fungal infections are also becoming supercharged on our warming planet thanks to climate change.

Paradoxically, an unlikely ally in this new era may be viruses, the zombie-like creatures of the microbial world. Despite their image problems, most viruses are quite harmless; some, possibly, quite beneficial to humanity. Read on. You might even come to admire the viral warriors of the biosphere.

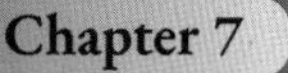

Chapter 7

Viruses That Do No Harm

In the unseen universe of microbes, two superpowers are locked in mortal combat: bacteria and viruses called bacteriophages. It's a brutal contest and the playing field pretty much runs through the entire biosphere. If there's a microbiological equivalent of the sci-fi thriller *Alien vs. Predator*, this is pretty much it.

Humanity has been a big beneficiary of this conflict. The creative destruction in this microrealm has powered molecular evolution and given rise to game-changing drug developments, not to mention gene-editing tools, that will transform modern medicine in the 2020s and beyond.

Consider bacteriophages. Phage is derived from the Greek word *phagein*, meaning to devour. These viruses are among the most abundant biological entities on Earth and absolutely dominate our oceans. If you stacked all of them end to end, they would extend out beyond the Earth's nearest 60 galaxies. Phages kill roughly 20 percent to 40 percent of the oceanic bacterial biomass every day, shaping the biogeochemical and nutrient cycles of the seas that cover 70 percent of the planet's surface.[1]

Like many viruses, phages first bind to the outer surface of bacteria and then inject their genetic information into their cell wall. Once inside, they commandeer the cellular machinery and multiply, eventually overwhelming and killing off their fellow microbes.

In response, bacteria have continuously evolved molecular mechanisms to avoid infection, only for phages to counter with adaptations of their own. Like Alice in Lewis Carroll's *Through the Looking-Glass*, both adversaries are running as fast as they can just to stay in place.

Leigh Van Valen, the renowned University of Chicago evolutionary biologist, had such pathogen-host struggles in mind in 1973, when he formulated his Red Queen hypothesis that species evolve and adapt not just for reproductive advantage, but also for their very survival.[2]

Our insights into this ultra-competitive microbial realm helped give rise to a gene-editing tool called CRISPR-Cas9, short for clustered regularly interspaced short palindromic repeats. (Cas9 is a type of protein.) That's not terribly poetic, but the scientific concepts behind the name are surreal, even a little awe-inspiring.

In nature's version, CRISPR is an ancient immune system strategy developed by prokaryotes like bacteria and archaea to neutralize phage attacks. It does so with a system that catalogs genetic information from attacking phages, then targets them for annihilation with specialized enzymes.

In one CRISPR microbial model that has been well-researched by scientists over the years, two RNA molecules and a Cas protein work together to unwind, cut and disable a phage's DNA to stop an infection dead in its tracks. CRISPR also compiles intelligence on its viral predators.

A CRISPR system copies and stores genetic information from the various phages with which a bacterium has had run-ins in the past. In other words, it codes a microbe's infection history with phages. Armed with that information, a bacteria or archaea can launch an effective counter-attack the moment it matches its archived genetic sequencing data to that of the attacking phage's DNA.

When these microorganisms detect the DNA of the phage, one of two things usually happens. If it's a known adversary, RNA molecules and a Cas protein (or similar enzyme) team up to attack and disable the intruder before a viral infection takes off. If the phage is new to the CRISPR system, a segment of the invader's viral DNA is copied and archived for a future counter-offensive.

Scientists have long wondered whether the CRISPR model, with its targeted and precise DNA knife-work, might someday be harnessed to edit live human cells. If so, disease-causing genetic mutations could be deleted. Life-enhancing genes might be added. Out of the labs of Jennifer Doudna and Emmanuelle Charpentier, came the CRISPR-Cas9 system based on research of the bacteria *Streptococcus pyogenes*.

In 2012, Doudna, a professor at the University of California, Berkeley, and Charpentier, who now works at the Max Planck Unit for the Science of Pathogens in Berlin, published a history-making piece of biotechnology research in *Science* that described the key components of the CRISPR model and showed how it could be refashioned into a gene-editing platform.[3]

With the CRISPR system for *S. pyogenes,* the Cas9 protein teams up with two RNA molecules to produce the biological equivalent of genetic scissors. One is named the CRISPR RNA, or crRNA, which carries the archived genetic information from previous phage attacks and targets the matching viral sequence.

The other is called a trans-activating RNA (tracrRNA), a molecule earlier discovered by Chartpentier, that essentially holds the crRNA with the Cas9 protein. The trio works together to find a sequence match in an attacking phage's DNA and then sever it, rendering it harmless.

Doudna and Charpentier showed how the whole system could be redesigned to direct Cas9 to cut specific sites in isolated DNA sequence segments of other cells. First, they swapped out the bacteria's crRNA with their own preferred genetic sequence and target DNA.

They also bioengineered a tracrRNA molecule to direct the scissors-wielding Cas9 protein to the target genetic sequence they wanted edited. Refinements by researchers since then have opened up a new world, with huge implications for drug research, agriculture and basic science.

It was a discovery that won them international renown and a Nobel Prize in 2020.[4] Yet the power to redesign life is also fraught with thorny dilemmas and implications for the future of the human species.

In 2018, the *MIT Technology Review* revealed that a Chinese scientist named He Jiankui had used a CRISPR system to edit the genomes of human embryos.[5] The specter of designer babies set off an international bioethical uproar. The scientist and his associates were later convicted of illegal medical practices by Chinese authorities and sentenced to prison terms.

Elsewhere, gene-editing technology has had a positive impact in both medical and agricultural science. In June, 2021, Intellia Therapeutics, a Boston biotech firm co-founded by Doudna, developed a treatment, using CRISPR gene editing, that for the first time could be directly infused into the bloodstream of a patient.

In previous clinical studies, human cells were removed, edited and then reinserted into test subjects. In this project, researchers used a CRISPR-Cas9 treatment to turn off a gene in liver cells linked to a debilitating and lethal condition called transthyretin amyloidosis, a disease that attacks the heart and nervous system. The research teams delivered the gene-editing package in nanoparticles made up of biomolecules that can be absorbed by the liver. All the patients in the trial experienced improvements.

Rise of the Superbugs

There's another Red Queen evolutionary dynamic at work between drug-resistant microorganisms and antibiotic agents. Phage therapy has the potential to be a powerful weapon in the escalating struggle against superbugs.

Twentieth-century discoveries of antibiotics, starting with penicillin in the late 1920s and streptomycin in the early 1940s, heralded a golden age for global healthcare. From 1950 to 1960, about half of the antibiotic drugs in common use today were discovered. Along with better sanitation and hygiene, antibiotics led to dramatic reductions in worldwide deaths from infections.

As a result, the Earth's biosphere has become inundated with antimicrobial molecules, not just antibiotics but also antivirals, antifungals and antiparasitics. Doctors worldwide have over-prescribed antibiotics, while industrial food producers have relied heavily on antimicrobial insecticides and fertilizers to boost food production and keep up with the escalating protein demands of an expanding global population.

Antibiotic consumption increased 91 percent on average globally, and 165 percent in low- and middle-income countries, between 2000 and 2015.[6]

Meanwhile, new classes of antibiotic drugs are getting harder to develop and only a handful of major pharmaceutical companies are investing aggressively in research and development to find new ones.

Facing massive selection pressure from this barrage of antibiotic drugs, microbes have developed remarkable biochemical defenses. Microorganisms have short lifespans yet are evolutionary sprinters, capable of mutating and passing along powerful genetic adaptations to the next generation.

What's even more impressive is their ability to swap genes amongst the same generation of microbes that aren't their direct offspring through a process called horizontal gene transfer. Bacteria, for instance, even those distantly related, transfer antibiotic resistance genes to each other, which has given rise to multi-drug resistant superbugs.

As a result, public health officials see a serious emerging health crisis. Some say it's already arrived. Drug-resistant strains of *Clostridium difficile,* which causes life-threatening diarrhea and colon inflammation, as well as new variants of gonorrhea and tuberculosis are considered urgent threats.

A multiple-drug evading fungus called *Candida auris,* which preys on people with weakened immune systems, is on a tear around the planet. The bloodstream infection is one of the world's most feared fungal microbes, often found lurking in critical-care units worldwide.

Researchers worry the COVID-19 pandemic, in which millions of sick patients have been treated with antibiotic drugs, may have supercharged the evolution of antimicrobial resistance. Some 700,000 people die annually from drug-resistant infections and related diseases, according to the World Health Organization, and that number could reach 10 million a year by 2050.[7]

Some scientists see phage therapy as an alternative or supplement to antibiotics in dealing with life-threatening bacterial infections. Whereas the number of potent antibiotic drugs have been dwindling since the 1940s, phages are omnipresent in the biosphere. Researchers did the math in 1999 and estimated that the number of phages was 10 to the 31st power. That's 10 followed by 31 zeroes. "All the world's a phage," they quipped.[8]

Phages are skilled molecular snipers. They take out their intended target without harming other microbes. Antibiotics, in contrast, are like indiscriminate fragment grenades that inflict a lot of collateral damage across the human microbiome, killing both good and bad microorganisms, causing side effects or even giving other microbial pathogens an opening to attack.

The specificity of phages, however, is also a potential downside. These viruses are very picky about which bacterium they attack and scientists have only started to skim the surface of understanding their characteristics. Also, in the West at least, phage therapy has taken a back seat to antibiotics.

While there are clinical trials underway in the U.S., getting approval for an experimental treatment like phage therapy has been a slow and cumbersome process, during which patients often die of complications from their infections before getting therapy.

Of the 785 U.S. patients that requested phage therapy between June, 2018, and May of 2020, only 17 actually received therapy and they waited an average of 178 days, according to data compiled by Keira Cohen, a pulmonary specialist and director of the Johns Hopkins Center for Nontuberculous Mycobacteria and Bronchiectasis in Baltimore, Maryland. "It's a little depressing," Cohen said in a talk to medical specialists. "That's little more than 2 percent."[9]

Phages to the Rescue

That said, interest in phage therapy is starting to grow, given the urgent health threat from drug-resistant, super-strains of bacteria and fungi that are killing patients. There's remarkable anecdotal evidence that phage therapy can bring patients back from the brink.

In 2019, the doctors and family of a gravely ill, 15-year-old cystic fibrosis patient named Isabelle Carnell-Holdaway reached out to Graham Hatfull, a leading phage researcher and biotech professor at the University of Pittsburgh.

Hatfull has spearheaded an effort to study and compile the genomes of mycobacteriophages — viruses that infect mycobacteria, a family of microbes that spawns diseases such as tuberculosis, which killed some 1.5 million people in 2020.

Isabelle was in very serious trouble after enduring a drug-resistant infection from *Mycobacterium abscessus* following a double lung transplant. She had been on immunosuppressive drugs at the time to prevent her body from rejecting the transplant.

Hatfull's lab was able to isolate the bacteria and came up with a phage cocktail while Isabelle was receiving palliative care and in grave condition. She is believed to be one of the first-ever patients to have received a phage treatment for mycobacterium, and in about six weeks the infection that had attacked her liver and skin was under control.

Steffanie Strathdee, a professor at the University of California San Diego's School of Medicine and Co-Director of the Center for Innovative Phage Applications and Therapeutics, learned first hand how effective phage therapy can be.

While vacationing in Egypt back in 2015, her husband, Tom Patterson, collapsed. It transpired that he had picked up an abdominal bacterial infection from a multidrug-resistant strain of *Acinetobacter baumannii.* As his condition worsened, no antibiotic, or combination of them, seemed to work for Patterson, a psychiatry professor at the same university in California.

As his health challenges turned life threatening, Patterson was transported first to Frankfurt, then back to his university's affiliated hospital in San Diego. He eventually fell into a coma and his prognosis looked grim.

Strathdee spent several frantic days reaching out to experts worldwide for possible treatments. Eventually, she and her colleagues were able to secure emergency approval from the U.S. Food and Drug Administration to treat Patterson intravenously with an experimental phage cocktail specifically designed to take out *A. baumannii*. It was a last resort treatment and fraught with risk.

"There were concerns that if you are injecting a billion phages into somebody's blood stream, they could undergo septic shock because the human immune system will see those phages as invaders," Strathdee told me. "We didn't know if the phages were going to cure him or kill him." Patterson responded to the treatment and eventually staged a full recovery. The Patterson's ordeal prompted both to recount their story in a book titled *The Perfect Predator*.

The story of phage research is one of the strangest in the annals of science. The bacterial destroying capabilities of phages were first observed in 1896 by the English chemist Ernest Hankin, who published works noticing that some sort of microorganisms seemed to be destroying the bacterium that causes cholera found in India's Jamuna and Ganges rivers, though he couldn't explain why.

Two decades later came the discovery of bacteriophages, which is primarily credited to Félix d'Hérelle, a brash and polarizing biologist, who worked at the Institut Pasteur in Paris where he researched dysentery. D'Hérelle gave these bacteria-killing viruses their name and published a paper in 1917 declaring they would transform infectious disease medicine.[10]

Comrade Phage

A devout Communist, d'Hérelle became a favorite of the Soviet scientific establishment and one of his books was dedicated to the strongman dictator Joseph Stalin. In 1934, he accepted a posting at a microbial research institute in Tbilisi founded by a Georgian microbiologist named George Eliava. The center, now called the George Eliava Institute of Bacteriophages, Microbiology and Virology, has emerged as one of the world's leading centers for therapeutic phage research.

Tragically for Eliava, he fell out of favor with Moscow and was executed in 1937, while d'Hérelle returned to Paris. In the West, the arrival of penicillin, the emerging geopolitical rivalry with the Soviet Union and skepticism about the quality of viral research generally resulted in phage treatment research being a backwater for decades.

With growing concern about the rise of superbugs, phage research has gathered renewed momentum. In 2018, the University of California launched its Center for Innovative Phage Applications and Therapeutics on its San Diego campus. Yale, Baylor and the University of Pittsburgh, as well as the Mayo Clinic, have major phage research programs underway.

There's a lot of research ahead to find, identify and isolate phage viruses to understand which bacteria they best match up against. "There are 10 million, trillion, trillion phages that are estimated to be on the planet and they come in all shapes and sizes," Strathdee points out.

Phage treatments have potential as alternatives or adjuncts to the dwindling number of effective antibiotics. Most of the infection-fighting drugs that have been introduced over the past three decades are variations of existing classes of antibiotics. Finding new ones can take more than a decade of research and $1 billion in capital to develop.[11]

Big drug companies facing investor pressure to deliver profitable results have moved away from the capital-intensive, low-return antibiotic market. There are an estimated 40 to 50 antibiotics in clinical development and governments are experimenting with new financial models to give Big Pharma and biotech companies more incentives to invest in antibiotic research.[12]

Phages attack bacteria by hijacking their replication machinery; using that power to develop antibacterial treatments may have potential advantages over antibiotics. The phage viruses attack specific bacterial strains, rather than indiscriminately eradicate any bacterium that gets in its way. Phages can also be effective against drug-resistant bacteria because they use a different molecular mechanism than antibiotics.[13] Finally, they can be combined with other classes of drugs to enhance treatment.

"It will never replace antibiotics, but I've seen many cases of synergy between antibiotics and phages," explained Strathdee. "Phages can actually be used to resensitize bacteria to antibiotics that they were previously resistant to. But there's quite a bit of basic science that needs to get done."

Phage cocktails for *Listeria monocytogenes,* the bacteria responsible for many food-borne illnesses, have received regulatory approval in the United States and Canada. America's NIH in 2021 awarded grants to support bacteriophage therapy research. These viruses are used in agriculture to protect tomato and pepper crops from pathogens such as *Pseudomonas syringae*.

However, there are several big challenges, both biological and economic, standing in the way of phage therapies. First, there's the obstacle of finding the right phage virus to attack the targeted bacterial infection.

So-called lytic phage viruses penetrate a bacterium, hijack its machinery and produce copies of itself until the cell explodes. However, lysogenic, or temperate, phages integrate into the bacterium and don't destroy the cell, or even go quiet. In some cases, they actually collude with bacterial cells to enhance their resistance to other phages and antibiotics.

Making things even more complicated, some "good" lytic phages can transform into "bad" lysogenic ones. Microbiologists need to better understand the risks of potential unintended consequences of this treatment.

Then there's the issue of creating the right economic incentives for drug companies to invest in phage research. Antibiotics, vaccines and other pharmaceuticals are compounds with proprietary and unique chemical structures that can be patented and monetized by drug companies.

Phages are living viruses in the natural world and courts worldwide have been reluctant to patent forms of life, or their constituent DNA and RNA. One potential way around this is for government research entities like the National Institutes of Health in the U.S. to underwrite the cost of creating a phage library that could be used for approved laboratories, hospitals and physicians.

Having publicly funded phage repositories might tempt more companies to invest and allow front-line clinicians to develop therapies much faster and possibly save more patients from life-threatening infections. They could then focus their research on proprietary methods to tweak phage genomes, synthetically engineer brand new phages or patent recipes for phage cocktails or combination drug therapies. Genetically altered phages could deliver cancer drug payloads to tumors, though that will take years of more research.[14]

Strathdee told me that she gets daily calls and emails from patients worldwide suffering from bacterial infections and seeking phage therapy as a last resort. "I've been constantly grieving for patients that we didn't reach in time," she said. The rise of superbugs is the next pandemic, according to Strathdee. "I see it on a daily basis, with somebody who steps on a nail and gets a bone infection that antibiotics can't touch. My husband was healthy one day, and the next day he was clinging to life."

Chapter 8

Terrestrial Distress Signals

Microbes have brought down empires. In the ancient world, soil erosion and land degradation were existential threats of mysterious origin, capable of triggering famines that ripped societies apart, even toppling complex civilizations. Consider the sudden population collapse of the ancient Mayan civilization in the lowlands of the Yucatán peninsula starting sometime in the ninth century.

At its peak, the broader Mayan empire numbered approximately 40 cities, many of which featured sophisticated water systems and managed landscapes that delivered robust harvests and generated wealth. The Maya produced hieroglyphic writing and symbolic artwork. Major centers such as Tikal and Calakmul, adorned with monument-festooned architecture, were believed to have had populations approaching 50,000.

Then something went terribly wrong. Some of the great Mayan city-states that spanned the Yucatán's southern lowlands in present-day Guatemala, as well as parts of Mexico, Belize and Honduras, turned into abandoned ruins in the course of roughly a century.

Peering back in time via paleoecological digs and soil chemical analysis, scientists envision a landscape under great stress from population pressure, deforestation and nutrient-depleted soil. The microbial systems that converted organic material and atmospheric gases into nutrients for crops and plants had unraveled.

"The ninth century collapse and abandonment of the Central Maya Lowlands in the Yucatán peninsular region were the result of complex human-environment interactions," two researchers wrote in a 2012 study.[1] "Large-scale Maya landscape alterations and demands placed on resources and ecosystem services generated high-stress environmental conditions that were amplified by increasing climatic aridity."

Climate models that recreate conditions in that part of the ancient world suggest that the large tracts of cleared land in the Mayan lowlands area increased surface temperatures and reduced precipitation. Analysis of stalagmites, a column of rock found in the region's caves, point to severe droughts, averaging 3 to 18 years in length, around the time of the mass exodus in the Yucatán.

Classical Mayan civilization took a massive hit from its exhausted soil. The Spanish conquistadors who arrived starting in the early 1500s and introduced smallpox eventually finished the job. History is replete with other examples of ancient societies — including the kingdom of Mesopotamia that cut across parts of current-day Turkey, Syria, Iraq and Iran — taking a savage blow from land ecology gone bad.

Our ancient forebearers didn't know of the existence of microbes or the role they play in soil health. Yet even in this current era of advanced soil science, precision farming and genetically engineered microbes and crops, the challenge of delivering protein to nearly eight billion people worldwide places enormous environmental stress on our lands.

The global population is expected to increase roughly to 9.7 billion by 2050 and nearly 11 billion by the end of the century, and farmlands already take up nearly 40 percent of the arable land surface of the planet.[2] Soil erosion remains a real and expanding challenge worldwide. Future agriculture land space is scarce and clearing new ground for food requires negative environmental tradeoffs such as more lost forest cover and more methane-producing livestock.

Add it all up, according to calculations by the Food and Agriculture Organization of the United Nations (FAO), and agriculture may need to produce 70 percent more food by mid-century while only being able to use 5 percent more land.[3] That means more chemical fertilizers that runoff into rivers and ocean systems. That means millions more livestock per year expelling climate-warming methane gas into the atmosphere.

Dirty Destiny

Natural soils are intricate menageries of microbial life. A handful of dirt is a microworld unto itself, teeming with tens of thousands of different microbial species. This consortium of bacteria, viruses, fungi and archaea communicates, collaborates and competes to process minerals, organic matter, nutrients and gases. It's a complex dance that shapes the planet's food web and atmospheric conditions.

Scientists have been aware of the transactional relationship between plants and microbes since the late 1800s, when Dutch microbiologist and botanist Martinus Beijerinck discovered a bacterial species living in the roots of legume plants such as soybeans and peas.

Beijerinck is also remembered as one of the founders of virology for identifying an infectious substance that attacked tobacco plants that he called "contagium vivum fluidum." He concluded that these were uniquely reproducing biological entities that were different from other organisms.

His work also showed how a certain type of bacteria, later classified as rhizobia, converted nitrogen from the atmosphere into a biochemical the plants can absorb to grow. The process is called nitrogen-fixing, and later research would reveal other transactional relationships in a broad symbiotic network of microbes and plants called the rhizosphere.

In this unseen microbial realm, plants secrete chemicals and nutrients beneficial to certain types of bacteria. Microbial bacteria convert hard-to-get chemicals useful to plant life and also protect against predatory viruses and parasites.

"There are a specific set of chemical signals that a plant sends out when it's stressed and it can exude sugars and amino acids at the same time to attract the bugs," according to Steve Maund. "It's a really beautiful biological system in a way." Maund is head of sustainability research and development with Syngenta AG, the Swiss seeds and pesticides heavyweight acquired by China National Chemical Corp., or ChemChina, in 2017 for $43 billion. The landmark deal was the largest-ever foreign corporate acquisition by a Chinese company.

Technological advances in the last decade have allowed scientists like Maund, who earned a PhD in ecotoxicology, to become interlocutors in that conversation. "There are opportunities to think about how to amplify those signals, make them more persistent or last longer," Maund told me. "You can actually spray plants to help them send the right signals to adjust to the changing conditions of the environment."

Those environmental pressures have been intense. Of the nearly 40 percent of the planet's land surface now devoted to agriculture, croplands take up one-third, while the rest consists of meadows and pastures for the grazing of livestock.[4] Plowing and overtilling have increased erosion of topsoil, a thin layer of organic-rich, dark spongy material formed from decomposed animal tissue and plants. That strata of soil plays a big role in containing water, nutrients and microbial life.

Soil can erode from natural causes such as wind and heavy rainfall. Yet the real culprit often is human activity, namely an expanding agricultural industry that's cleared forests for more farmland, allowed livestock to overgraze and engaged in other unsustainable farming practices.

It's a lesson humanity keeps relearning. America's decade-long Dust Bowl in the Midwest and Southern Great Plains during the 1930s compounded the economic suffering of a global depression. Over-farming and the massive loss of water-preserving and microbial-rich grasslands set the stage for an economic catastrophe when a severe drought hit, triggering massive dust storms starting in 1931. Millions of acres of cultivated land were rendered useless for farming and an area the size of Texas lost much of its topsoil.

"I shall never forget the fields of wheat, so blasted by heat that they cannot be harvested," American President Franklin Delano Roosevelt told his nation in a radio broadcast in 1936 upon returning from a trip to the drought-stricken region. "I shall never forget field after field of corn stunted, earless stripped of leaves, for what the sun left the grasshoppers took."[5]

Land degradation, or the loss of soil's organic matter, fertility and microbial diversity, is no less a pressing environmental and national security concern in the 2020s. Soil resources in East African countries such as Burundi, Kenya, Rwanda, Uganda and Tanzania are being lost to erosion, threatening the food and water security of millions of citizens in the region.

An estimated 25 percent of the world's total land has been degraded and 24 billion tons of fertile soil are being lost every year.[6] One study found that one-third of the top soil had been eroded in the corn belt farmlands in the central United States.[7] If current trends continue, as much as 95 percent of the Earth's land area could have low soil fertility by mid-century.[8]

The driest parts of the planet tend to be in developing countries, whose populations are most at risk from land degradation and drought. An estimated 3.2 billion people may face malnutrition and forced migration under worst-case future scenarios, according to the Intergovernmental Science-Policy Platform on Biodiversity and Ecosystem Services.[9]

The widespread conversion of land with complex ecosystems such as forests, grasslands and peatlands, that have the capacity to absorb and store carbon, could worsen the concentrations of carbon dioxide and nitrogen oxide in the atmosphere.

There's scientific evidence (which we'll explore in a later chapter) that vast terrestrial carbon storage regions called sinks, such as the Amazon rainforests and permafrost zones in higher latitudes, risk becoming net emitters of greenhouse gas because of shifts in microbial behavior. Some microbiologists worry that global warming may speed up the metabolism of microbial populations that produce CO_2 and methane.

When cattle belch or pass gas, they release methane into the atmosphere. Yet even some basic crops such as rice, a crucial food source for half the world's population, has a carbon footprint comparable to that of international aviation. Rice grows in flooded paddies blocking oxygen from seeping into the soil — an ideal setting for bacteria to grow, feed off plant matter and release methane gas.

Calling All Microbes

Humanity isn't powerless to reverse these trends, but it will take a multi-disciplinary effort and continued advances in microbial science to shift farming practices in a more sustainable direction.

Senior executives running the world's biggest food and agricultural companies are facing a heat blast of pressure from investors and consumers who are increasingly aware of the role the sector plays in climate change. A destabilized planet isn't good for business it turns out.

The organic food industry that sells products free of additives and industrial chemicals is no longer the fringe movement it was in the 1980s and 1990s. In 2021, organic food revenue was expected to hit $221 billion and $380 billion by 2025.[10]

Then there's the sheer firepower of the sustainable investment sector. Financial assets focused on environmental, social and governance (ESG) strategies are growing 15 percent a year and are on track to reach $53 trillion by 2025, up from $22.8 trillion in 2016, according to forecasts by Bloomberg Intelligence.[11]

Some skeptics dismiss the ESG movement as marketing gimmickry and empty virtue signaling by financial institutions and companies. There's doubtlessly some of that at work. Yet it's also true that chief executive officers who blithely ignore sustainability in their business practices may not remain in the corner office for long. Companies are under pressure from institutional investors and consumers to come up with credible plans to make sure their operations and supply chains aren't degrading farm lands or destroying the microbial biodiversity the planet needs to recycle greenhouse gases.

An activist fund called Engine No. 1 shocked the business world in 2021 by winning a shareholder proxy battle with powerful ExxonMobil, America's biggest oil company. It was able to obtain three board seats for executives with clean energy backgrounds by convincing a broad swathe of Exxon shareholders that it was in their economic self-interest to do so. Another fund, New York-based Third Point, has called for the breakup of Royal Dutch Shell.

That said, there are no quick fixes in sight to heal our lands. The majority of the world's expanding middle-class population isn't going to suddenly prefer synthetic meat grown in a lab over sirloin steak that comes out of a slaughterhouse to reduce the ecological strain of the massive livestock population worldwide.

Global meat consumption keeps rising year after year, as incomes grow in low- and middle-income countries. On current trends, the food industry, which generates about 37 percent of annual greenhouse gases, could see a 30 percent to 40 percent increase in emissions by 2050.[12]

Turning the tide will take a multifaceted and self-reinforcing strategy that embraces more carbon-neutral styles of farming and food production, reduces the environmental hoof-print from livestock and develops new protein sources that will gradually diversify our diets away from heavy meat consumption. Advances in microbial science will be essential to all three goals.

Microbial Multitudes

Starting in the early 2010s, big agrochemical multinationals like Bayer, BASF, and Monsanto began acquiring microbial biotech firms. Just as with human microbiome research, agricultural science researchers have leveraged new sequencing technologies to explore the complex interactions of microorganisms underneath the soil.

In the 1970s, agricultural scientists still used microscopes to study microorganisms one species at a time. Now, they have access to rapid-fire DNA sequencers to study vast colonies of them and can genetically modify microbes to optimize certain characteristics or functions in the soil.

Brian Brazeau, former president of North American operations of Danish industrial biotech firm Novozymes A/S headquartered near Copenhagen, thinks the technology is there to improve the quality of soils and transform industrial farming.

Novozymes develops and manufactures industrial enzymes and microorganisms for agricultural, food, biofuels and household applications. Among its 2020 product rollouts were a biocleaning product called Microvia that uses bacteria to clean hard surfaces such as kitchen countertops and Protana Prime that brings out the natural umami flavor from plant proteins for alternative meat products. "We are now at the point where we can understand the biology well enough to engineer and to deploy it in very specific ways," according to Brazeau.

Our planetary health depends a great deal on reducing the chemical fertilizers, greenhouse gas emissions and infectious disease risk that flow out of modern-day industrial farming. "Biology did a great job for millions of years, and it didn't take us very long to step in and disturb that," according to Brazeau. "You don't change any of the challenges that the world is facing without looking at more sustainable means of food production."

AgBiome, based in the Durham, North Carolina, Research Triangle Park, has isolated and sequenced more than 100,000 microbial strains worldwide and created a genomic database. Microbial genes have been screened for their performance against pests. New gene-editing tools like CRISPR-Cas9 and synthetic biology techniques make it possible to take genes from one microbe and insert them into another.

AgBiome and BASF have developed a biological fungicide based on a modified bacteria called *Pseudomonas chlororaphis* that protects vegetables from diseases with far less of an environmental impact than traditional products. Soil microbes are being developed to help crops like wheat, pepper and corn better withstand drought conditions by stimulating their roots to more efficiently absorb water even from drying soil.

Next-generation microbial products are coming out of the labs and being manufactured by big companies like BASF, Bayer and Syngenta. According to Ian Jepson, a researcher, molecular geneticist and head of the technology development at Syngenta's Innovation Center at Research Triangle Park in North Carolina, biologically-derived agriculture products are a $4 billion global market that's expected to hit $10 billion by 2030, as the European Union and other governments push for reductions in the use of crop protection chemical products. Gene-edited agricultural crop varieties and biologicals are being developed to address farmers' needs that were previously handled by chemical products.

One of the more interesting examples of the new biocontrols being developed, Jepson points out, involves a molecule called RNAi, as in RNA interference, which can be engineered to target pests and pathogens by disabling their genes. "We identify a target in the pest organism that is essential for life, and you inhibit it by using an RNAi molecule. It's highly specific to the pest," says Jepson.

It's unlikely that biologicals and new seed varieties will completely supplant chemical products, but we may be able to use far less of these traditional plant control and growth products that indiscriminately damage microbial networks in the soil or end up as toxic runoff in waterways.

The big challenge going forward is thoroughly understanding how microbial consortia work together in the wild to make sure biological products actually deliver the results they promise. "A lot of companies are pushing biological products with no clue how they work," explained Jepson. "They are in spray and pray mode."

Brazeau, formerly of Novozymes, also thinks there are still basic research issues that need to be sorted out. "What we are missing is an understanding of how to tease apart, predict and design entire communities of living organisms," he reckoned. "Bacteria are just like people: throw 1,000 bacteria together and it's very difficult to predict and design what they're going to do." Further out into the future, we may be able to redesign how entire microbial ecosystems work. "The Holy Grail is nitrogen fixation," said Brazeau. "Making the bacteria and plants more efficient at consuming the nitrogen that's in the environment prevents that nitrogen runoff and stops a lot of these big algal blooms and big dead zones in the ocean."

Tweaking the microbiomes on our farmlands to absorb more nitrous oxide, a greenhouse gas with far more heat-trapping power than carbon oxide, could also help mitigate global warming. Improving soil productivity would be another huge benefit, given growing concerns in some countries about food security. It certainly is the case in China, which is home to about 7 percent of the world's arable farmland but roughly a fifth of the global population.

The Chinese Government signed off on the massive Syngenta acquisition, in part, to boost its self-sufficiency in agriculture. China now represents about 20 percent of the company's global revenue. Syngenta has launched a digital farming program that helps mainland farmers precisely monitor crop performance, fertilizer use and other metrics for optimal crop yields.

Syngenta's chemical crop protection, bioherbicides and genetically modified seeds could help China expand its food production more sustainably as it tries to stay apace with the expanding demand for protein as it becomes more wealthy and its middle class swells.

In the West, big food companies, meanwhile, are promoting regenerative farming practices among their suppliers. General Mills, the U.S.-based consumer products multinational, has pledged to reduce greenhouse gas emissions across its value chain from "farm to fork lift" by 30 percent by the end of the decade from 2020 levels and achieve zero emission levels by 2050.[13]

General Mills also aims to improve agricultural practices on one million acres of farmland in its supply chain. That means using microbes and other natural ecosystems for nutrient recycling, improving biodiversity in the soil, and employing rotational grazing land and covering crops on empty fields. It has established pilot programs with oat growers in the Northern Plain states and wheat farmers in Kansas.

It's not enough that companies are setting carbon reduction targets for their own operations. Benchmarks also need to be extended to suppliers lower down in the production chain. Third-party organizations such as the Science Based Targets initiative (SBTi), a partnership backed by climate groups like, CDP Worldwide, United Nations Global Compact, World Resources Institute and the World Wild Fund for Nature, are helping companies set up achievable standards and a system to verify progress.

When *Bloomberg Green* analyzed the climate pledges of 187 different global companies with a 2020 deadline, 138 had hit their marks. Some of those goals were quite modest, but this level of commitment from the world's biggest companies didn't exist at the start of this century.[14]

Microbial Chefs

Food technology has evolved dramatically over the last two decades, with the potential to change global diets later in the century. If we choose to take a leap, microbes might be crucial allies. Humans weren't always ravenous carnivores. Our proto-human ancestors started to diversify their diets beyond grass, flowers, fruits, barks and insects to protein-rich meat and marrow roughly 2.6 million years ago, based on fossil records showing primitive butchery marks on ancient animal bones.[15]

It would take a real optimist wearing green-tinted Ray-Bans to believe that meat will be eliminated from the human diet anytime soon. Shifting even a portion of humanity away from deeply ingrained dietary behavior will be a massive undertaking. Equally as daunting will be convincing modern-day consumers that microbes are part of the solution.

Yet whether we acknowledge it or not, microbes have been integral to our diets for much of recorded human history. Fermented foods that rely on yeast include staples like beer, wine, bread, yogurt, cheese and pickled vegetables. Now, microbial science is being enlisted to make alternative meats with an authentic taste amid growing environmental concerns about the impact of expanding livestock populations and ethical qualms about slaughtering animals in factory farms.

In 2009, Patrick Brown, a professor of biochemistry at the Stanford University School of Medicine, started to think about the ecological impact of raising livestock for global meat production. With an estimated 19 billion chickens, 1.5 billion cattle and 1 billion sheep and pigs, there are many more farm animals than humans living on the planet.[16]

Supporting that population is land intensive, given that farmers need to house them but also find space to grow feed crops. Forests have been cleared to make room for these creatures destined for slaughter and consumption. That's reduced global biodiversity. While they live, livestock produce pollutants that contaminate soil, water and air.

So Brown launched Impossible Foods Inc. with the idea of making an appealing and commercially viable meat alternative made of plants. Brown has taken on complex problems in the past. Together with colleagues at Stanford, he developed a technology called DNA microarrays that monitor the activity of genes in a genome. He's also credited with valuable work in deciphering gene expression patterns to classify cancers.

You'd think making an artificial hamburger would seem trivial by comparison. Not for Brown, who has described recreating the molecular interplay that makes meat delicious as one of the most important scientific questions in the world right now.

"Arguably, the most important and urgent problems our species has ever faced are the catastrophic meltdown in biodiversity and the relentless progression of climate change, both of which the use of animals in the food system is a major player," Brown said in a podcast with the American Society of Microbiologists.[17]

Impossible Foods is growing quickly and sports a multi-billion dollar valuation, so he's had some success cracking the code. He got there with the help of two microbes.

Early on, the research team at Impossible Foods figured out that what makes meat appealing to humans is its high concentration of heme, a family of iron-containing molecules. Iron is an important source of nutrition, but heme is also an important chemical trigger during cooking.

"Heme catalyzes very specific types of chemical reactions," according to Brown, "that transform abundant, simple nutrients into this explosion of hundreds of diverse volatile odorant molecules. When you experience them together, they add up unmistakably to the smell and taste of meat."

As the company searched for an appealing recipe for a meat replacement, its scientists needed another source of heme. His research team found one by exploring the relationship of legumes plants and their symbiotic bacteria called rhizobia that convert nitrogen into an energy source plants can absorb.

It turned out that two bacterial species, *Bradyrhizobium* and *Sinorhizobium*, produced a heme protein that could be used for alternative meat production. The company's researchers then genetically engineered and fermented certain types of yeast that, in combination with the heme from soybeans, creates a biochemical called leghemoglobin that's been instrumental in creating a convincing ground beef imitation.

Plant-based meat products use far less land, water, fertilizer and labor to make. They're healthier, too. Consumers will shift if alternative meats are satisfying to eat and are price competitive. Brown, for one, sees a distant future in which cattle are viewed as pets, not the driving force of the world's food system.

That day seems a way off, but investors are waking up to the commercial potential of alternative protein sources. The alt-meat category has turned very competitive with new players like Impossible Foods and Beyond Meat squaring off against established players like Kellogg and Conagra Brands in the vegan chicken category.

In 2020, some $3.1 billion in investment flowed into companies focused on alternatives to animal-based foods such as plant-based meat, egg and dairy products, fermentation technologies and lab-grown meats, according to a survey by The Good Food Institute.[18]

Fresh from the Labs

In December 2020, the Singapore Food Agency approved the world's first synthetic meat product for sale. Soon after, a restaurant in the city-state called 1880 sold a cell-cultured chicken product designed by a San Francisco biotech company called Eat Just, whose investors include the estate of the late Microsoft co-founder Paul G. Allen and wealth funds backed by Qatar.[19]

Lab-grown meat is genuine animal meat with the same taste, texture and smell. Instead of slaughtering an animal, the food is made from cell types that recreate a livestock's biological structure. Since Dutch scientist Mark Post created the first synthetic burger in 2013, research money has poured into this emerging technology. The trick will be producing cell-cultured meats on an industrial scale that's affordable and safe.

The idea of eating meat made in a lab may strike some as unappetizing. Then again, one could argue that devouring the skin of an animal carcass is rather weird, too. With cultured meat, the essential stem cells of animals are collected then grown in bioreactors, where they are fed nutrients and supplemented with proteins and growth-enhancing microorganisms.

The cells then develop into the skeletal fat, muscle and connective tissue that constitute meat. The whole process can take weeks to complete. Commercial production doesn't harm the environment, kill animals or require heavy antibiotics use.

Animal food production is a force multiplier for antimicrobial resistance, with an estimated 70 percent to 80 percent of all antibiotics worldwide used to keep farm animals healthy, that is before we send them merrily into the slaughterhouses.[20] Cell-cultured meats could be a $140 billion market by 2030, accounting for about 10 percent of the global meat industry, according to an analysis by investment bank Barclays.[21]

Further out into the horizon are new approaches to microbial fermentation. Microbes are complex entities that have a diversity of biosynthetic pathways that produce flavors and aromas that can reproduce the attributes of meat and dairy products.

These microorganisms can also be sequenced and re-engineered genetically to produce a desired product. They can then be produced in huge quantities and affordably through fermentation, in which enzymes from raw ingredients change a microorganism's molecular characteristics. Yeast enzymes, for instance, convert sugar and starches into alcohol.

Another bioconversion strategy aims to tackle two environmental problems simultaneously: reducing food waste and land use devoted to growing livestock and feed. A third of all food goes to waste, resulting in $1 trillion in economic losses.[22]

Enter a Cambridge University spinoff called Better Origin. The start-up has come up with an innovative way to convert that refuse into nutrient-rich animal feed. First, the unused food is grinded and processed to remove pathogenic microbes. Then the waste is routed into a portable AI-monitored bioconversion unit and fed to the larvae of black soldier flies.

Inside these units, biosensors monitor temperature and CO_2 levels, while cameras and computer algorithms make decisions to promote optimal growing conditions. It's a highly-autonomous bit of high-tech farming. Before the larvae hatch, they're turned into nutrient-rich feed for chickens.

"Food waste doesn't exist in nature; it's an entirely man-made problem," Better Origin CEO and co-founder Fotis Fotiadis told me. "We take this natural concept of insect bioconversion and we blend it with cutting-edge technology. Then we reintroduce it back into the food chain by turning food waste into high-value animal feed."

It's not as crazy as it might seem. There's a reason that many types of poultry have evolved to hunt and forage for insects, particularly live larvae whose microbial profile improves the digestive system of chickens, making them more productive and active.

"The protein from insects has a richer amino acid profile and is more nutritious than plants," explained Fotiadis. "You end up populating their gut flora with good microbes, and the microbiome of the hen is much stronger than it would be without the insects."

The nutritional quality of larvae from black soldier flies, locusts, grasshoppers and crickets make them ideal candidates to replace soymeal and fishmeal in the diets of poultry, pigs, cattle and fish, according to a study by the United Nations Food and Agriculture Organization.[23]

Depending on the species, insect feed could be swapped out for 25 percent to 100 percent of livestock feed that requires huge tracts of land, water and fertilizer to sustain. It's potentially an elegant solution that would take some much-needed pressure off our land and the terrestrial microbiome we need to secure our food security and make sure our soil retains as much carbon as possible.

If microbes have shown us anything, it is this: symbiotic relationships between the microbial world and humanity exist throughout nature. New ones are possible, maybe even crucial. Nowhere is that truer than in our oceans that take up 70 percent of Earth's surface. They need to heal in the decades ahead. Microbes can help, if we let them.

Chapter 9

Microbial Ocean Lords

The Baltic Sea is a geological newcomer among Earth's oceanic water basins, carved from the last ice age's receding glaciers only some 11,000 years ago. Yet there's nothing particularly youthful about the Baltic Sea in the 2020s. Marine scientists fear it's at risk of future ecological peril, given its pollution-powered microbial imbalances, overfishing, and the impact of climate change.

This sea is an inland arm of the North Atlantic Ocean, extending from Southern Denmark to the northern reaches of the Scandinavian Peninsula near the Arctic Circle. It's a relatively shallow and enclosed body of water encircled by nine countries — Estonia, Denmark, Finland, Germany, Latvia, Lithuania, Poland, Russia and Sweden — with varying levels of development and environmental protections.

Surrounded by roughly 85 million coastal citizens, heavy industry and agricultural regions, the sea has been inundated with sewage and chemical fertilizer run-off from regional river systems for decades. A buildup of nitrogen and phosphorus has altered the microbial profile and biochemistry of these waters in deleterious ways.

The Baltic Sea is home to some of the world's biggest dead zones, large bodies of oxygen-starved water that are stressing marine life. The Baltic cod and herring populations have crashed in recent years and the sea is one of the world's fastest-warming.[1]

If all that weren't enough, more than 75 years ago the victorious allies in World War II thought it was a good idea to dump 65,000 tonnes of chemical weapons and related agents produced by Nazi-era Germany into the Baltic.[2]

Those drums and shells of hazardous waste are now at risk of leaking. In 2013, researchers at Poland's Military University of Technology discovered traces of mustard gas on the sea bed in the Gulf of Gdansk.[3]

A more immediate problem is the damage the sea is experiencing from an ecosystem that's gone off the rails. In our oceans, microorganisms called phytoplankton, ranging from microalgae to bacteria, are the starting point in several marine food webs.

Phytoplankton tend to float in the upper waters of the ocean, require sunlight to survive and feed off of inorganic nutrients such as nitrates, phosphates and sulfur. They also play a central role in the recycling of carbon dioxide, nitrogen and oxygen to and from the atmosphere.

One type of phytoplankton called cyanobacteria is an ancient microbe and a biological innovator, responsible for arguably the most profound transformation in the early Earth's environment that opened an evolutionary pathway to complex biological life.

During much of the Archean Eon that started with the formation of the planet's crust about four billion years ago, the atmosphere was virtually oxygen-free. It consisted primarily of gases like hydrogen sulfide, methane, and carbon dioxide that spewed from volcanoes.

Blue-green cyanobacteria flourished in primordial oceans. Eventually, they evolved the ability to harness energy from sunlight, carbon dioxide and water, a process called photosynthesis, all of which created oxygen as a byproduct. Scientists now believe, based on ancient stone formations, fossil records and retrospective genetic analysis, that cyanobacteria triggered the Great Oxygenation Event about 2.4 billion years ago, when O_2 started to accumulate in the atmosphere at meaningful levels.[4]

Today, oxygen represents about 21 percent of the atmosphere. Photosynthesizing land plants and forests later came on the scene and also became big oxygen producers as well. Yet even to this day, oceanic phytoplankton, with its assortment of bacteria, algae and drifting plants, generate about half of Earth's oxygen in our biosphere.

One species of bacteria called *Prochlorococcus* is among the smallest photosynthetic organisms on Earth, yet produces an estimated 20 percent of our biosphere's oxygen, more than all of the world's tropical rainforests.[5] Amazingly, this bacterial overachiever wasn't discovered until the late

1980s, thanks to the work of Sallie Watson "Penny" Chisholm, a biological oceanographer at the Massachusetts Institute of Technology and a National Medal of Science winner, and a team of research collaborators.[6]

When it comes to the Baltic Sea, oxygen is in short supply. The agricultural run-off, sewage and stormwater flowing into these waters have carried with it nitrogen and phosphorus. Microbial bacteria and algae feast on these nutrients and, under the right conditions, reproduce at such a furious rate that they form sprawling phytoplankton blooms that can extend hundreds of kilometers.

Viewed from space satellites, the spirals and vortexes of blue-green algae blooms are strangely alluring. Up close, they can be quite dangerous, releasing toxins that are lethal to marine life, shellfish, domestic animals and humans that come in contact. These microbial biomass formations at the surface of the water unbalance the ocean food web, block sunlight and can trigger mass kills of fish, sea plants and other microbes.

At some point, the feast ends and the blooms die off and sink, but the resulting breakdown of microalgae or cyanobacteria, depending on the type of biomass, creates a boomerang effect. Bacteria move in to devour the organic material, creating a massive stench. As the bacterial population explodes, they deplete oxygen levels within the water.

Decades of fertilizer run-off and sewage have reduced oxygen levels along the coastal waters of the Baltic to lows not seen in some 1,500 years, researchers concluded in a 2018 study.[7] The sea's cumulative dead zone in recent years has expanded to about the size of Ireland.

The Baltic Sea may be a preview of what's to come if we don't figure out smarter strategies to heal our oceans by mid-century, scientists warn. The number of dead zones, including massive ones in the Arabian Sea and Gulf of Mexico, has quadrupled since 1950.

Trouble Beneath the Waves

Then there's the possibility of unwelcome feedback loops emerging as climate change warms the oceans. Climate scientists are warily tracking the Atlantic Meridional Overturning Circulation (AMOC), a global-scale oceanic conveyor belt that includes the Gulf Stream System, that has been weakening in recent decades.

The massive current system transports warm surface water from the equator northward, while pulling cold deep water back south. All told, the current moves nearly 20 million cubic meters of water per second, almost a hundred times the flow rate of the Amazon River, according to oceanographer and climatologist Stefan Rahmstorf from the Potsdam Institute for Climate Impact Research (PIK), initiator of a study published in *Nature Geoscience* in early 2021 that garnered international media attention.[8]

Marine microbes hitchhike around the globe on this hydrodynamic superhighway. Scientists are trying to decipher how the warmer ocean water and environmental shifts affect microbial populations and their ability to recycle greenhouse gases.[9]

Warmer and more acidic oceans have been linked to the spread of pathogenic, water-borne microbes such as *Vibrio cholerae*, the bacterium that causes the acute diarrhoeal infection of the same name. The public tends to associate cholera with contaminated water and sewage. However, these bacteria also prosper in warm coastal waters of moderate salinity in the tropics. In cholera-prone regions along the Pacific Coast of Latin America and the Bay of Bengal, heavy concentrations of *V. cholerae* have been linked to plankton blooms and warmer currents.[10]

In 2016, microbiologist and renowned cholera researcher Rita Colwell and colleagues measured *Vibrio* populations in samples compiled by the Continuous Plankton Recorder (CPR) survey, which has been doing research in the Atlantic Ocean since the early 1930s.

By comparing samples with historical sea-surface temperature fluctuations, the team found that *Vibrio* populations jump and infections rise when oceans warm. In the future, the team concluded, there's a risk of cholera infections in new regions such as northern Europe.[11]

Then there's the dreaded *Vibrio vulnificus*, which can cause death or serious illness to humans exposed to contaminated raw shellfish or whose open wounds come in contact with the pathogen. The flesh-eating bacteria is antibiotic-resistant and sometimes requires limb amputation to treat. About 20 percent of people with the skin infection variety of the disease die, sometimes within a day or two of infection, according to the Centers for Disease Control and Prevention.[12]

Vibrio species tend to congregate in estuaries along coastlines. However, researchers believe that increased temperatures and rising sea levels will shift the geographic range of pathogenic bacteria. It is likely the roaming grounds of *V. vulnificus* will increase as water temperature rises and salty seawater extends further into formerly freshwater reaches of coastal rivers, scientists predict.[13]

Microbial Web

The far bigger research task will be developing models that can give the world some predictive power about what climate change will mean for the ocean's complex and shape-shifting microbiome later this century. In the early 2020s, there are major blind spots in our understanding of these microorganisms.

This much we do know: Earth's early history experienced very destructive swings in microbial balances in our oceans and things didn't end well. Past extinction events have coincided with warm climates and oxygen-deficient oceans.

Worldwide, oceans have lost around 2 percent of their dissolved oxygen since the 1950s and are projected to lose 3 percent to 4 percent more by the year 2100.[14] That may not sound like much, but even modest reductions can ripple through the marine biosphere. Much of the oxygen loss is concentrated in the upper 1,000 meters of water, where biodiversity is most abundant.

Already, coastal regions experiencing historically low levels of oxygen are increasing. If humanity continues to use the oceans as a dumping ground for fertilizer run-off, sewage and animal waste, by the end of this century more than half of the world's marine species may face serious risk of extinction.[15]

One way or another, microbes will have a big say as to whether that kind of dystopian scenario in our seas awaits us or not. It has only been in the last two decades that scientists have started to fully appreciate how integral these infinitesimal organisms are to ocean health and the climate.

In 2010, researchers from eighty countries announced the results of a decade-long, $650 million project called the *Census of Marine Life*.[16] It found that more than 90 percent of the biomass of the ocean consisted of an invisible realm of microbes: viral particles, bacteria and other single-cell organisms. They existed in a variety of settings, from coral systems and marine life to hydrothermal vents and the deepest trenches of the ocean.

Single-cell phytoplankton float at the top of the ocean, where they convert sunlight into chemical energy, consume carbon dioxide and release oxygen. In turn, they and other microorganisms called ciliates, single-cell creatures with hairlike arms used for movement and food gathering, are consumed by tiny animals like crustaceans and zooplankton.

Bigger fish prey on smaller ones as we move up the food chain to apex predators such as tuna, squid, and sharks, as well as marine mammals like seals, dolphins and whales. As marine life produces waste, ages out and dies off, microbes move in and clean things up, resetting the entire food chain cycle. Without microbes, the ocean's food web and entire ecosystem would surely collapse.

Over great spans of time, microbes have also developed symbiotic relationships with other marine life. Corals release nutrients to attract certain types of bacteria that return the favor by supplying nitrogen, decomposing waste and warding off harmful microbes.

Bacteria supply neurotoxins to blowfish and octopuses, sea stars and horseshoe crabs to evade predators. Migratory sea mammals like humpback whales have complex skin bacteria that protect them from disease-causing pathogens and may improve swimming speeds by limiting the accumulation of barnacles.

Deep Blue Carbon Pumps

Microbes play an essential role in the ability of our oceans to capture and store about a third of the man-made carbon dioxide emissions and much of the excess heat generated by the world's greenhouse gas emissions. One of the most urgent scientific questions in the decades ahead is whether climate change will disrupt these carbon cycling mechanisms and result in our oceans being less effective at absorbing greenhouse gases.

The ocean is the world's biggest carbon storage system with two major pumps: a biological one is related to the oceanic food web and the other is powered by the chemistry of the ocean.

The biological pump involves a series of processes powered by photosynthesizing phytoplankton at the ocean's surface that soak up carbon dioxide from the atmosphere. Some are consumed by other microbes and marine animals that recycle carbon in the food web. Other phytoplankton die off and convert carbon from organic to mineralized form, some of which falls to the ocean floor and is locked away for years.

Carbon dioxide can also dissolve directly into seawater, creating carbonic acid. This happens extensively in the higher latitudes of the polar regions, hence the climate change worries about the rapid warming dynamics in the Arctic. Dense water currents in colder regions drag down the dissolved carbon toward the seabed.

Over the centuries, the oceans have stored massive amounts of carbon, perhaps 50 times more than is currently in the atmosphere. As concerning as the climate challenges on the horizon are (by 2025 CO_2 in the atmosphere will be at levels not seen since 15 million years ago) things might be even more dire without the carbon absorption from our oceans.[17]

Before the world's industrial revolution and the burning of fossil fuels, the absorption rate of CO_2 between the atmosphere and the oceans appeared fairly well balanced. That may no longer be the case in the decades and centuries ahead.

Since the dawn of the industrial era, the oceans have taken in an estimated 525 billion tons of carbon dioxide.[18] The conversion of CO_2 into carbonic acid has made the oceans more acidic over the past 200 years, which is detrimental to the health of various shellfish such as oysters and mussels, complex reef systems and various fish species that may have trouble adjusting to the chemical changes in our seas.

Then there's the impact that rising temperatures will have on microbial behavior and carbon capture. Warmer water may speed up the metabolism of microbes and the pace of chemical reactions and genetic mutations. Rising ocean temperatures have coincided with pathogens that have ripped through marine life such as the 2013–2015 occurrence of a mysterious sea star wasting syndrome.[19]

Will the ocean's role as a carbon sponge be compromised? Despite their tiny stature, microbes are fairly mobile and adept at finding new habitats when confronted with environmental challenges. Marine microbiologists are watching for shifts in the population dynamics of phytoplankton, given their crucial role in oxygen production and carbon dioxide recycling.

With their massive numbers, distribution over a large portion of Earth's oceanic surface area, and rapid reproduction rate, marine phytoplankton have some built-in advantages over fixed terrestrial plants that photosynthesize to ride out dramatic shifts in ocean water temperature.

A few studies have raised the alarm about a marked decrease in phytoplankton populations, though there's far from a scientific consensus about that.[20] What's still missing are long-term data sets by which to judge fluctuations in the density of these microbes. Marine microbiologists are the first to admit that we're still stumbling around in the dark.

Global microbial ocean expeditions are gathering samples to do metagenomic analysis to get a sense of the population trends over long time spans. However, it will take more research in the years ahead to gain an accurate read on the future direction of phytoplankton communities and what that means for global greenhouse gas emissions.

Another potential risk is the future stability of methane hydrates, ice-like formations of the greenhouse gas in the deep sediment layers of the ocean floor. It's a massive, untapped energy source and, according to the United States Geological Survey, one of the plentiful hydrocarbon sources in the world.

This fiery ice is locked up in hard-to-reach ocean depths and permafrost zones, so energy companies haven't yet been able to commercialize this resource fully. Japanese companies, nevertheless, are making strides in extracting gas from methane hybrid reserves in a seabed of the Nankai Trough, off the eastern coast of Japan's main island of Honshu.

In recent years, researchers have surmised there's a low risk of sustained global warming destabilizing the methane hydrates and unleashing massive amounts of the greenhouse gas into the atmosphere. The depth of the gas hydrates in sediments would make it difficult for the methane bubbles at the seafloor to make it to the surface and then the atmosphere in most cases, a 2016 study concluded.[21]

The issue of hydrates aside, researchers have found evidence of methane seepage off the Pacific Northwest coast of the United States and in the Arctic Ocean. In 2020, scientists also discovered a methane leak in the sea floor in Antarctica.[22] The good news is that methane-eating microbes capture much of the released gas that bubbles up within our oceans. The less good news is that scientists aren't sure that will be the case in the decades ahead if the planet and oceans continue to warm.

More broadly, the big unknown in marine microbial science is precisely how global warming will shape migration and population patterns of the microorganisms that recycle greenhouse gases.

Researchers at the University of Southern California, working with the National Oceanic and Atmospheric Administration and the University of Edinburgh, have developed a model of how billions of microorganisms might possibly adapt to warming ocean currents. In 2020, it was made freely available to scientists worldwide.[23]

In its simulations of future climate scenarios, some microbial species were able to adapt to changes in their environment quickly. Others saw their populations suppressed but evolved new genetic traits that allowed them to recover over the long run when temperatures had locked in at higher levels.

Efforts also are proceeding apace to better catalog the vast marine microbial world through investigative studies and expeditions. Millions of genes, and about 1,000 entire genomes, have been identified in new lineages of microbes by the Global Ocean Sampling Expedition of the J. Craig Venter Institute.

A consortium of Australian universities and research institutes found more than 175,000 unique species of microbes along Australia's coastline and in the Southern Ocean.[24] Off the coast of Maui, Hawaii, in the central Pacific, scientists have used a fleet of robots to collect samples of ocean microbial communities for genome sequencing. The quest to map out the marine microbial universe will be a multi-decade effort.

Saving the Big Blue Sea

Healing our oceans and reducing the destructive impact of climate change is one of the most complex challenges ever to face humanity. The abundance and biodiversity of marine life have steadily diminished since the 1980s. One-third of the major fish stocks are overfished and coastal wetlands that protected against soil erosion, supported biodiversity and sequestered carbon have been lost to coastal development. One in four species of sharks, rays and skates are at risk of extinction. By 2050, there may be more plastic than fish in our oceans.[25]

If there's an upside, it may be that world governments have finally started to grasp that the ocean is a finite resource and its size and depth won't prevent a reckoning if we continue unsustainable fishing practices and treat our seas as a waste disposal system.

Encouragingly, the number of marine protected areas in the national waters of individual countries has been steadily growing over the past two decades. These conservation zones are a buffer against industrial fishing and other environmentally disruptive activities that cover nearly 8 percent of the world's oceans.[26]

The deep ocean regions beyond national borders represent 43 percent of the Earth's surface and they are the most significant repository of marine and microbial biodiversity. International management has been ineffectual due to a mish-mash of regulatory schemes.

Calls are growing by scientists and environmental groups to cordon off 30 percent of our oceans in conservation zones by 2030 to stabilize the biodiversity of marine and microbial life in our seas. In late 2021, 196 governments began to negotiate a new set of global biodiversity targets for the world's oceans and land in a push to stabilize ecosystems.

Talks are underway to create a binding, multilateral treaty under the United Nations Convention on the Law of the Sea to improve the world's environmental stewardship of the high seas, the vast part of the ocean that falls beyond national borders. The proposed treaty would make it easier to set up a system of marine sanctuaries in the deep seas and dictate how industries can operate in the shared waters.

One of the thorniest issues is ownership of genetic resources from microbes and marine life with potential business or pharmaceutical value. If these deep waters are a common resource owned by all nations, wouldn't it be a form of biopiracy for advanced economies or a handful of powerful companies to reap commercial gains from their discoveries in these vast seas?

Programs such as the Global Fishing Watch are using satellite radar imaging to track ships suspected of illegal fishing in national waters. We will need a much more extensive surveillance system to monitor deep-sea sanctuaries. An international research team has launched a program to strap sensors onto albatrosses in the southern part of the Indian Ocean to track bad behavior.

Repairing the damage to marine life will take multiple generations and an interlocking set of mutually-reinforcing policies. Such policies will need to aim at stabilizing global greenhouse emission targets, promoting renewable energy, embracing regenerative farming and biofertilizers, and expanding aqua-farming as well as the development of plant-based and lab-grown seafood products.

Microbial Ocean Healers

At the same time, interesting research is underway to explore ways microbes might help with the cleanup effort by harnessing their bioremediation talents.

The Great Pacific Garbage Patch has gyres of floating debris that extend from the West Coast of America to Japan, much of it in the form of plastic fragments from grocery bags, water bottles, personal care products or remnants from washing synthetic clothing that ends up in the oceans.

Plastic packaging is strong, light and inexpensive, but from an environmental perspective it has a tragic design flaw. Drink and food plastic packaging is typically made from polyethylene terephthalate (PET) and designed for single use and disposal. Yet the underlying material can last for centuries.

If we had full-proof systems in place to collect, recycle and repurpose these materials, the planet wouldn't be awash in plastic. Unfortunately, about one-third of plastic products produced annually aren't collected and spill out into the environment, often the sea.

On top of that, much of plastic is made up of petrochemicals that contribute to global warming, and major oil and chemical companies continue to spend billions building out the plastics infrastructure worldwide.

In our oceans, microplastics have become part of the food web, showing up in the stomachs of bird species and marine life. The chemicals in these plastics incorporate other toxins floating in the water, which further undermines marine health.

This a solvable problem, yet will require a global approach to increase rates of reuse, collection and recycling. Ecologists argue that we also need to reconsider our "throw away" consumer culture and accelerate research of bioplastics and bacterial remediation technologies.

Companies have every incentive to rethink their packaging strategies. Not only does survey data show consumers demanding it, but some environmental groups and political leaders in the United States and Europe are weighing policies that would hold consumer goods and food companies, rather than taxpayers, responsible for paying for the waste collection, management and recycling of plastics.

Over the last decade, research into bioplastics made from natural, renewable feedstocks such as corn starch, vegetable oil, sugar cane, and algae have gained considerable research and media attention.

Manufacturing bioplastic creates far less pollution than making regular plastic out of petrochemicals. In the mid-2010s, Coca-Cola launched PlantBottle, a PET container partially made with Brazilian sugarcane, to reduce its carbon footprint.

Yet this is far from a comprehensive solution. Making bioplastics is still very expensive, and diverting crops to make packaging imposes new pressures on food production and land use. Green plastics can be recycled and are compostable, meaning that microbes can decompose them into natural materials that blend harmlessly into the soil of a landfill.

However, not all bioplastics are fully biodegradable. If non-oil-based plastic ends up in the oceans, it may still blight the waters and kill fish. It's going to take time for this technology to develop and become more commercially viable. Other strategies will be needed as well. One might be unleashing bacteria directly on ocean pollution.

Nature's Tiny Pollution-Busters

In 2010, an explosion ripped through the Deepwater Horizon oil rig, operated by a drilling company called Transocean and leased by British oil colossus BP Plc., about 41 miles off the coast of Louisiana.

The accident killed 11 workers and resulted in the biggest spill in the oil industry's history, dumping the equivalent of 4.9 million barrels of oil and 250,000 metric tons of natural gas into the Gulf of Mexico over nearly three months. The disaster contaminated the open ocean and 1,300 miles of the Gulf's shoreline.

A U.S. Congressional review of the disaster later that year estimated that roughly 41 percent of the spill had been directly recovered, burned, skimmed or chemically dispersed. The balance of the oil wasn't precisely accounted for, falling under categories such as "naturally dispersed" (13 percent), "evaporated or dissolved" (24 percent) and "other" (22 percent).

The Deepwater Horizon calamity took place in an era in which marine biologists had access to metagenomic and bioinformatic tools to track the unseen swarming colonies of microbes and study the ecological impact of the disaster in detail.

A retrospective analysis of the spill published in 2020 by the Gulf of Mexico Research Initiative revealed how the world's oldest and tiniest creatures helped rescue the Gulf of Mexico from an epic man-made environmental catastrophe.

As one of the world's biggest offshore, oil-drilling regions, researchers suspected that microbes in these waters had been primed to consume hydrocarbons, given years of small leaks from drilling operations.

When the Deepwater rig blew, oil-consuming microorganisms moved toward the spill, flourished off the new food source and quickly expanded into blooms. After they died off, other microorganisms moved in. Eventually, hydrocarbon-degrading microbes accounted for as much as 90 percent of the microbial community around the disaster site.[27]

Biodegradation also occurred in low-oxygen environments such as the deep seafloor and shoreline salt marshes. Different species of microbes swapped genetic and biochemical processes to keep the cleanup going over the years. Even the chemical dispersants used in the initial phase of the cleanup, which are toxic to marine life, were degraded by bacteria able to pull apart sulfur-containing compounds, according to DNA sequencing data compiled by the researchers.

There's no denying that the Deepwater spill delivered incredible damage to the environment. Some marine species are still recovering. Nor is the Gulf of Mexico anybody's idea of pristine waters. Its low-oxygen dead zone is estimated to be one of the world's largest. Yet many were surprised by the bioremediation skills of these tiny creatures.

Could bacteria be engineered to identify and clean up the estimated five trillion pieces of microplastics that pose health hazards to marine life and humans? Japanese researchers have found a newly discovered bacterium called *Ideonella sakaiensis* that secretes enzymes that are effective at degrading PET.[28]

Microbial Sea Harvest

As our awareness of the vastness of ocean microbes increases, so do the opportunities to discover natural chemicals and bioactive marine compounds to create anti-cancer agents and new treatments to fight bacterial, viral and fungal infections.

Since biological life began in the oceans billions of years ago, the marine environment has been home to a rich diversity of microorganisms. Gene editing has opened up a pathway to enhance the microbial genes that are beneficial to human health. Marine microbes have been around longer in evolutionary terms, are far more diverse and have developed an array of bioactive compounds not found in land-based microorganisms. We've only begun to tap that potential.

Winding back to the 1950s, marine biochemists have appreciated the pharmaceutical properties of sea sponges that have developed symbiotic networks with microbes.[29] Sponges are immobile and lack natural defenses against fishes and turtles. In alliances with microbes, they've developed interesting chemical defenses.

Scientists were able to isolate chemicals from a shallow-water sponge common in the Caribbean which have been useful in cancer and antiviral drugs, including Zidovudine, or AZT, which was a breakthrough drug in AIDS treatment. In 2020, a team of researchers at the University of Wisconsin–Madison, made a significant finding by studying the microbiome of the humble sea squirt, an ancient and solitary marine filter feeder.[30]

They found an antifungal compound, named turbinmicin, that destroys drug-resistant strains of the fungus *Candida auris*, which has been spreading around the globe and preying on humans with weakened immune systems. The compound has performed well in early animal trials and could end up saving millions of lives.

An experimental Alzheimer's drug called Bryostatin is based on a chemical compound found in the symbiotic microbes of marine animals known as *Bugula neritina* that researchers believe might help restore neurodegenerative damage caused by the disease. The results in early trials have been mixed, but the drug continues to attract serious interest from researchers and is still on the FDA-approval track.

Meanwhile, microbial fermentation technologies show long-term potential in developing alternative seafood products that will give consumers another option to industrially fished products or those grown in acqua farms. A food tech company called Aqua Cultured Foods has developed a way to make fish protein with a fermentation technology using fungi.

It's still early days, but companies are starting to see the outlines of a new ocean economy. Marine biotechnology is already a multi-billion dollar business as companies discover new ways to make food, fuel and materials from bioactive compounds in our seas.

Yet we will need to heal our oceans before this potential can be fully realized. If we continue to trash the seas, the microbial backlash in terms of carbon release, food production and human health could cost humanity dearly.

Chapter 10

Tiny Sky Pilots in the Jetstream

On March 15, 2021, Beijingers woke up to a dystopian world of suffocating smog and a blue-orange sky. A witch's brew of pollutants and dust from vehicular exhaust and coal-fired plants engulfed the Chinese capital. On top of that, a sandstorm had blown in from the Gobi Desert of Inner Mongolia.

International media outlets called the miasma an "airpocalypse" health emergency. Less appreciated, though, was the fact that the bad air above Beijing was almost certainly teeming with microbial life that had hitched a ride on the pollutants.

China's toxic air has been an emotionally charged political issue since the early 2000s. On that particular March day in Beijing, a measure of a fine particulate matter called PM 2.5 that can reach deep into a human respiratory tract reached levels considered extremely hazardous by groups like the World Air Quality Index Project.[1]

The week before, at China's annual National People's Congress legislative meeting in Beijing during which the city was also submerged in bad air, Communist Party leaders pledged to remove heavy pollution by 2035.

In China, lung cancer is the leading cause, some 20 percent, of all cancer deaths. The country's mortality rate from lung disease is expected to increase 40 percent by 2030 over 2015 levels.[2] Air pollution kills around seven million people globally a year, according to the World Health Organization estimates, and siphons away $5 trillion from the global economy.[3]

Chinese scientists have discovered links between the country's world-class air pollution and the spread of microbes that are highly resistant to antibiotics.[4] Widespread circulation of drug-resistant bacteria is scarcely unique to China. Yet if we end up in a post-antibiotic world in which routine infections may be life threatening as most experts fear, China may be among the countries worst hit.

Like anywhere else in the world, the Chinese atmosphere is home to a complex microbiome of tiny life forms. Microorganisms are swept up from farm lands, water treatment centers, urban population centers, and aquatic systems, then redistributed over great distances. Researchers are trying to better understand how airborne viruses, bacteria and fungi called bioaerosols are moving around as climate change shifts atmospheric circulation.

Evidence suggests that microorganisms transported by air currents may be interacting and swapping genes, influencing weather patterns, and posing possible new infectious disease threats in the decades ahead.

With its 1.4 billion population, massive livestock industry, and rapid economic growth, Chinese consume far more antimicrobial drugs than the world average.[5] Just as in Western countries, Chinese doctors have misprescribed antibiotics designed for bacterial infections for viral ones like influenza or common colds. That, in turn, circulates more drug-resistant genes into the environment.

Ordinary Chinese have little trouble securing antibiotics from the local pharmacy without a prescription, even though the practice is illegal.[6] Antibiotic use in the livestock industry, particularly in rural China, is even higher than human consumption levels.

Farmers liberally feed the drug to cattle, swine, and chickens to keep them healthy before they're dispatched to the slaughterhouse. As a result, antimicrobial genes are finding their way into meat products and the water supply from livestock waste. A 2015 study revealed that 58 percent of surveyed Chinese schoolchildren tested positive for antimicrobials.[7]

Researchers have detected and sequenced great amounts of antibiotic resistance genes in hospitals, inland surface water, soil, sewage treatment, plant effluents, and animal waste, all of which has coincided with a rise in cases of Chinese suffering from hard-to-treat bacterial infections. In 2017, some 73,000 people in China developed multi-drug resistant tuberculosis, representing 13 percent of the global cases, according to the World Health Organization.[8]

These potentially dangerous drug-resistant microbial genes also hover in the skies above Beijing. In 2019, Chinese microbiologists collected indoor and outdoor air samples during polluted and clean air days.[9] They isolated and genetically sequenced the microbes and then compared the results with a database of antibiotic resistant genes.

What they disclosed was concerning. Bacterial concentration is consistently higher in polluted weather, particularly with a type of microbe called gram-negative bacteria (GNB). The genes of GNB tend to cause infections of wounds and the bloodstream, as well as conditions such pneumonia and meningitis. Not only are they stubbornly resistant to most available antibiotics, they also can swap genetic information that helps other bacteria do so as well.

High Life

Aerobiologists have only started to grasp the breadth of microbial air dynamics over the last several decades. It's been long understood that a spinning Earth and atmospheric circulation patterns create tradewinds that sweep up dust and sand, and redistribute them around the world.

The American plant pathologist Fred Campbell Meier coined the word "aerobiology" in the 1930s, when he researched the relationship between plant disease and airborne fungal spores. In 1935, Charles Lindbergh and Meier flew over the Arctic and made atmospheric collections of microbes.[10]

Yet the idea that a complex network of living organisms might reside in the lower troposphere of the atmosphere, or reach the stratosphere, seemed far-fetched until this century.

In 2006, two environmental microbiologists, Christina Kellogg and Dale Griffin with the U.S. Geological Survey, published a paper proposing a microbial "air bridge" hypothesis, in which global masses of dust and sand carried microorganisms around the planet.[11] The continents regularly exchanged bacteria, viruses and fungi as part of a microbiome of microorganisms in the skies above us.

Consider that a random teaspoon of soil roughly holds up to a billion bacteria. In dry parts of the world, desert winds aerosolize and transport several billion tons of soil that are swept up in tradewinds that flow from continent to continent. The microbes swirling in the air currents are passengers in an intricate aerial highway network.

"Huge dust events create an atmospheric bridge over land and sea, and the microbiota contained within them could impact downwind ecosystems," Kellogg and Griffin speculated. "Such dispersal is of interest because of the possible health effects of allergens and pathogens that might be carried with the dust."

The evidence that bacteria can use Earth's atmospheric currents to fly thousands of miles through the air has been documented by genetic analysis of microbes worldwide. Bacteria isolated very far apart geographically have been discovered to have the same genes in common.

In a 2019 study, scientists collected and sequenced *Thermus thermophilus* bacteria from Mount Vesuvius and other parts of Italy, as well as in Russia and Chile.[12] The samples were scalding hot, about 160 degrees Fahrenheit, and were thus unlikely to have been transported by birds or via human travelers.

For comparative purposes, researchers traced genetic information the bacteria stored from their encounter with bacteriophages, those predatory viral microbes whose behavior inspired CRISPR gene-editing tools.

They expected to find little similarity, assuming the bacteria would have evolved independently. Instead, they found examples of identical pieces of DNA. That supports the idea that there's a planet-wide mechanism that transports microbes and that the bacteria that come in close contact in the air may be sharing genetic information by swapping genes.

Early on, aerobiology tended to focus on the air dynamics of allergens like pollen that trigger allergic reactions in some people. Today, it's a multidisciplinary field that uses molecular research techniques, as well as gene sequencing tools, to track microbes and identify biorisk in both the open air and indoor environments like intensive care units, meat packing plants and schools.

Stratospheric Microbes

David J. Smith, director of the Aerobiology Laboratory with NASA's Ames Research Center in Silicon Valley, has been intrigued with microbial life since his undergraduate days at Princeton University. He joined NASA in 2007 after earning a doctorate in biology and astrobiology at the University of Washington.

"I'm absolutely fascinated by the interconnectivity of the systems on our planet and the idea that we have these atmospheric bridges that can make the world seem a lot smaller. What's happening on one continent can influence a downwind continent," Smith explained to me.

"Borders don't exist in the atmosphere," he added. "Whether it's wildlife, smoke crossing borders, or dust from the Sahara Desert coming across the Atlantic ocean to the Eastern Seaboard of the United States, by the way carrying airborne microbes, we are all connected by the atmosphere."

Smith has come up with novel approaches to capture a collection of microbes with aircraft flying at high altitudes. Microbes have been found 36 kilometers, or 22 miles, up in the stratosphere above the planet's protective ozone layer, where it's cold, dry and ultraviolet radiation is lethal to most forms of life.

In the lower troposphere stage of our atmosphere, microbes regularly jump between continents on massive dust plumes visible from space satellites, live in cloud formations and even affect weather patterns in a process called bioprecipitation.

"It looks like anywhere between 5 to 50 percent of aerosols larger than 2.5 micrometers that we sample in the atmosphere are biological," Smith said. "If you were to take a milliliter of cloud water on average, there would be 10,000 fungal cells and 100,000 bacterial cells. That's truly stunning and we didn't know that until relatively recently."

These tiny sky pilots also are integral to weather patterns through a phenomenon known as ice nucleation. Contrary to the common assumption, pure water doesn't automatically freeze at 0 degrees Celsius (32 degrees Fahrenheit). It can super cool and remain in liquid form at temperatures far below that.

As Smith points out, to freeze at higher temperatures water needs an impurity or tiny particle, around which water molecules can align geometrically and crystallize into ice. Once this nucleation happens, the water will freeze and then fall to the Earth's surface as rain, sleet or snow depending on the time of year.

It turns out that microbes, and particularly bacteria, are very efficient at helping cloud water freeze. "Bacteria are accelerating the precipitation process when it comes to physics," according to Smith. "In a sense, they can induce their own precipitation."

A plant pathogen called *Pseudomonas syringae* is particularly skilled at ice nucleation and has been incorporated into artificial snow products used by ski resorts. Researchers aren't entirely sure, but one theory is that bacteria may have evolved this talent as a way to get a return ticket back down to the Earth's surface aboard rain droplets or snow crystals.

Smith is attuned to the idea that humans are bottom dwellers living underneath an ocean of atmosphere. One of the key facets of his work is trying to figure out the upper boundaries of Earth's biosphere. In other words, how high do you need to go up before you lose the planet's biosignature?

To assess how airborne microbial patterns work at high altitudes, Smith's team flew a NASA C-20A aircraft over the U.S. Sierra Nevada mountain range on two consecutive days in June of 2018.

The aircraft, a military version of a Gulfstream business jet redesigned for science experiments, featured a bioaerosol catching device Smith's team had engineered. The team flew at four different altitudes ranging from 10,000 feet to 40,000 feet.[13]

They aimed to track changes in microbial populations across the vertical stages of the troposphere, the lowest and cloud-populated layer of the atmosphere that extends up to 10 kilometers (6.2 miles or 33,000 feet) above sea level. Smith's team collected quite a bit of microbial biomass, enough to reliably analyze with shotgun, metagenomic analysis what's up in the skies.

They found a wide array of common microbial species, many of which probably originated from California's Central Valley agricultural region and regional waste treatment facilities. Smith's team also found a strong DNA presence of a marine microbe called *Tetrahymena* during its flight at 10,000 feet. The protozoan microorganism has never been detected at such heights in previous survey work.

The team also picked up traces of airborne-pathogens such as *Alcaligenes*, a bacteria found in the guts of vertebrates that can cause opportunistic infections, and *Mycobacterium* that triggers pulmonary disease. Also circulating in the atmosphere was a fungal-like microbe called *Pseudoperonospora cubensis* that can infect cantaloupes, cucumbers, pumpkins and watermelons.

Unmanned, experimental high-altitude balloons have collected bioaerosols as high as 125,000 feet in the stratosphere, the second layer of the atmosphere. Microbes have been found even in the dry and ultra violet-drenched regions above the Earth's protective ozone layer.

How did they get up there? Smith and others speculate that violent upward air convections from thunderstorms, hurricanes, or perhaps volcano eruptions, give microbes the escape velocity they need to break into the stratosphere, though very few microbes survive in those conditions. Some species of bacteria or archaea, so-called extremophiles, as we shall see later, have evolved DNA repair capabilities to deal with high-radiation environments and even survive in open space.

Understanding how bioaerosols migrate and disperse in the air is crucial to tracking infectious disease spread. In the Southwest of the U.S., outbreaks of Valley Fever have been linked to pathogenic microbes transported by dust.[14]

In sub-Saharan Africa, desert storms are thought by scientists to carry a deadly bacterium that causes meningococcal meningitis, a serious infection in the lining of the brain and spinal cord. Outbreaks have been so common that public health officials have dubbed the region of 26 countries and 300 million people the African "meningitis belt."[15]

Microbial Air Attacks

Aerobiologists are also worried about pathogens closer to the ground and in closed environments like office towers and schools. The airborne transmission of novel pathogens that our immune systems don't recognize is the nightmare scenario for infectious disease experts.

Our lungs offer disease-causing microbes the most efficient route into our bodies. While ingested microorganisms must travel through an acidic environment before they can colonize tissue, inhaled bioaerosals can land unencumbered on the moist surface of the respiratory tract and start reproducing.

In the current pandemic, there has been widespread debate about to what degree COVID-19 is an airborne threat. At issue, is whether the virus is transmitted via respiratory droplets which tend to drop to the ground quickly and within about six feet, or longer-lasting aerosols that can linger in the air for hours and migrate around an indoor setting depending on the ventilation system.

Early on in the COVID-19 pandemic, the Centers for Disease Control and Prevention in the U.S. emphasized the droplets dynamic over aerosols and that shaped its early advice to some degree about mask wearing and the risk of people gathering indoors.

However, in 2020 a group of infectious disease experts in an open published letter pointed to numerous studies that demonstrated "beyond any reasonable doubt that viruses are released during exhalation, talking, and coughing in microdroplets small enough to remain aloft in air and pose a risk of exposure" up to 10 meters (32 feet).[16] One study showed that SARS-CoV-2 can linger in the air and remain infectious for up to 16 hours in aerosol form.[17]

By the end of 2021, research revealed that a variant of the airborne COVID virus called Delta that originated in India was considered one of the most contagious seen by scientists in the history of infectious disease. People infected with Delta carried 1,000 times more virus in their nasal passage than those infected with the original strain.[18] Then came another variant called Omicron, which was even more transmissible than Delta.

The air-transmitted *Mycobacterium tuberculosis,* another highly transmissible pathogen, is a bacterial infection that attacks the lungs, brain, kidneys and spine. Infected individuals spread it when they cough, talk, laugh and sneeze, and it can quickly rip through overcrowded cities and confined spaces like airplane cabins.

Tuberculosis is an old nemesis that has been treatable with vaccines. However, new strains are proving very resistant to the most potent TB drugs: isoniazid and rifampin. Some 1.5 million died of TB-related illnesses in 2020.[19]

One of the big microbial science and infectious disease research priorities is understanding the rise of fungal infections and interplay, if any, with climate change. Fungi run from the complex eukaryotic organisms we can see to invisible particles of molds and yeasts that are classified as microorganisms. Big or small, many types of fungi thrive in warmer and humid weather conditions, while the spores of some species become airborne and spread geographically. As the climate changes, fungi are finding new homes around the world.

A certain group of fungi produce toxic compounds called mycotoxins that grow on agricultural foodstuffs such as cereals, fruits, nuts and spices. These toxins can make their way into the food chain via livestock feed and can be dangerous to animal and human health.

Fungi species such as *Aspergillus flavus* produce a toxic chemical called aflatoxin, prevalent in warmer regions, that can contaminate crops in the field and during storage. These toxins have been linked to liver cancer.[20] The former Iraqi dictator Saddam Hussein's Iraqi regime explored the use of aflatoxin in his country's biological weapons program between 1985 and 1991.

One fungal species roaming around the globe is *Candida auris,* which was first identified as an ear infection in 2009.[21] It has since spread worldwide and is very resistant to existing antifungal agents. *Candida auris* can enter the body through open wounds, the bloodstream or the ear canal and tends to attack people with compromised immune systems.

It's very lethal when it finds its way into hospital settings, particularly intensive care units where patients have breathing issues, feeding tubes and are using a lot of antibiotics. Researchers are still trying to figure out whether this fungal species has evolved independently and, if not, how it has spread so quickly globally. It's still very much a mystery.

Microbial SkyNet

In the anxious days after the September 11, 2001, terrorist attack in the U.S., letters laced with anthrax began appearing addressed to news organizations and American senators. Five died and seventeen were sickened after breathing in spores of the bacteria, in the worst biological attack in American history.

A government taskforce, consisting of investigators from the Federal Bureau of Investigation, the U.S. Postal Inspection Service and other law enforcement agencies spent years trying to crack the case. In 2008, the Department of Justice and the FBI were preparing charges against Bruce Ivins, a microbiologist who had worked with anthrax at the U.S. Army biodefense laboratory at Fort Detrick, Maryland. Ivins committed suicide before the criminal indictment was released. The FBI formally closed the case in 2010.[22]

We face potential airborne threats both from aerosolized biological agents dispersed by terrorists and by nature that pose catastrophic risk to human health, agriculture and the economy. Government leaders and scientists are exploring ways to create a global air surveillance network that can monitor dangerous microbes for building forecasting models and responding quickly to threats. Yet coming up with an effective system that's reliable has proven a technological challenge.

Designing biodetection technology is far more complex than tracking chemical, radiological, or nuclear agents, which have specific molecular structures. In contrast, microbial threats from terrorists, or Mother Nature, could come from viruses, bacteria or toxins and can mutate quickly.

In 2003, America's Department of Homeland Security (DHS) started the BioWatch program, designed to provide early indication of an aerosolized biological weapon. The federal program set up aerosol collectors around the country and a system that could evaluate samples in about 36 hours.

The program has been criticized by scientists and congressional researchers over the years for being too slow and using outmoded technology.[23] Air samples, for instance, had to be physically transported by car to labs for analysis and federal reviews have found instances of false positive readings of biological agents.

BioWatch has been replaced with a new fully-automated program called Biological Detection for the 21st Century (BD21). The new system, expected to be operational in the mid-2020s, uses artificial intelligence to quickly detect and analyze traces of biological agents of concern.[24]

Biosensors that pick up suspicious particulate readings in the air will be automatically routed to a computerized analysis network. Algorithms will sift through a database of genetically sequenced microorganisms looking for anomalies, while machine learning programs will use mathematical models based on historical data to make a decision or prediction about the new substance under scrutiny.

Air samples will be acquired by a field team sent to the site of the suspicious biological agent and routed to a lab, where scientists and counter-terrorism officials can conduct tests to amplify the DNA of the agent to determine its health risk, and whether it has been engineered by terrorists and possibly its point of origin. The government believes BD21 will be able to detect and identify airborne agents in as little as four hours.

Scientists foresee climate change leading to even higher numbers of microbes entering the atmosphere. Africa's Sahara Desert, the world's largest, has expanded by about 10 percent since 1920, resulting in more dust and bioaerosols circulating in the atmosphere, according to a study by the National Science Foundation.[25]

Deforestation and land reclamation for farming to feed an expanding global population is also kicking up more soil-based microorganisms into the atmosphere. New weather patterns with higher microbial concentrations may have unpredictable consequences regarding the spread of infectious diseases and allergens.

Getting better situational awareness of shifting microbial and atmospheric aerosol data would give governments and public health officials a chance of identifying an emerging avian virus outbreak or bioterrorist attack early. It will never be a fail-safe system, yet we definitely have the sensor, AI and data analytical technologies to better manage man-made and natural microbial risk if we choose to.

There's also scientific value in studying how microbes survive and interact in the air, particularly at high altitudes. Unlocking the secrets of airborne microbes may have implications for drug development, biogeochemical cycling and weather management. The radiation-resistant genes evolved by some microbes might have application for biotechnology and long-distance space travel.

If microbes play a role in the ecology of clouds, might we use that knowledge to better predict the weather or even induce rain in the world's most parched regions? Can airborne bacteria degrade pollution or even absorb carbon emissions? The atmosphere is an unexplored microbiological frontier. Important discoveries may await if we look skyward.

Chapter 11

Why Microbes Control Our Climate Destiny

Arctic ecologist Susan Natali has spent much of her professional career trekking through icy stretches of Alaska and Siberia, analyzing the impact of climate change on the shape-shifting landscape of the remotest places on the planet.

As a senior scientist and Arctic Program Director at the Woodwell Climate Research Center in Falmouth, Massachusetts, Natali studies the chaotic forces now reconfiguring permafrost, the icy sublayers of soil whose fate may alter climate change scenarios later in the century.

In the summer of 2020, a heat wave swept through Siberia, with temperatures reaching 38 degrees Celsius (100.4 degrees Fahrenheit), the highest ever recorded reading within the Arctic Circle.[1] A series of wildfires in these polar regions set a record for fire-generated carbon dioxide emissions, as they were on track to do again in 2021.[2]

In recent years, permafrost in Russia, the U.S. state of Alaska, northwest Canada, and elsewhere has been thawing, fracturing, and buckling. Massive sinkholes and gullies have appeared in permafrost zones as the underground ice that holds the soil together liquifies.

In Russia, scientists have reported the appearance of massive craters from methane gas explosions on the Yamal Peninsula in Northwestern Siberia, bounded by the Kara Sea and Gulf of Ob. An eruption believed to be linked with pressurized methane gas underneath rapidly thawing permafrost during the 2020 heat wave created a 164 foot-deep (50 meters) hole, according to Russian scientists.[3]

Ancient pathogenic bacteria and viruses may be welling up from the deep as well. In 2016, an anthrax outbreak ripped through reindeer populations in the Yamalo-Nenets region of Siberia as a heatwave melted permafrost and possibly exposed an infected reindeer carcass to the environment.[4] Some 90 people were sent to hospital and one young boy died. The government dispatched troops trained in biological warfare to help with the emergency.

International governments have committed to holding down the global temperature increase to well below 2 degrees Celsius, ideally beneath 1.5 degrees Celsius, this century compared to pre-industrial levels. The Arctic region has already smashed through that threshold, and the cycle of rapid warming underway is expected to push that figure even higher mid-century, according to models by climate scientists.

In a 2021 study, Natali and her research colleagues at Woodwell and Harvard University described a process of sudden thawing events that reconfigured the landscape. The team also chronicled the climate damage from a run of wildfires in the Arctic.[5] "Fire-induced permafrost thaw and the subsequent decomposition of previously frozen organic matter may be a dominant source of Arctic carbon emissions during the coming decades," the team concluded.

These sudden thaw holes in the ground, and even the creation of entire new lakes, are changing the risk assessment of permafrost carbon emissions. Yet they aren't yet fully accounted for in most Earth system models used in the United Nations Intergovernmental Panel on Climate Change's (IPCC) assessments, considered the gold-standard in the field and that guide world leaders on climate policy.

Prior to the pandemic, Natali and her teams had been regularly conducting research in both Alaska and Siberia, usually from May to September, with a few trips during the winter months. Every year that they returned to their field research sites, they were amazed by remarkable transformations in the landscape.

"I had been spending a lot of time in the Yukon-Kuskokwim Delta in Southwest Alaska, which is a peatland area with permafrost that's rapidly thawing," Natali told me. "The fires in the area had made the ground unstable because the peat moss had been burned off. There were places where you'd walk and your legs would fall into the ground. It's really incredible to see the land just split open."

When Natali conducted ecological research near Chersky in Northeast Siberia, where massive ice wedges underneath the permafrost have collapsed and melted, she saw newly protruding animal bones in the soil dating back to the last ice age and beyond.

These thawing permafrost zones are experiencing newly created landforms, known as thermokarst, that are exposing ancient permafrost, and the frozen microbes contained within, to warm air for the first time in millennia. Entire lakes are being swallowed up by landslides and erosion, while entirely new water bodies are being created by the structural realignment underneath the soil.

Thermokarst lakes arise as underground ice turns to liquid water, the land surface slumps and creates a depression that then fills with ground ice water, rain, and snow melt. The water underneath and on the edges of these spontaneous lakes, in turn, speeds up the thawing of the frozen soil around its perimeter. As a result, the thermokarst lakes tend to keep expanding in size and depth.

In 2018, a NASA research team found 72 locations, across 11 thermokarst lakes in Alaska and Siberia, that are leaking methane; the team used radiocarbon dating to determine the age of samples. They also compared emissions to normal lakes where gradual permafrost thawing is taking place.[6]

Field tests show methane gas is bubbling up from the depths of both new and older lakes in Alaska and Siberia. That's concerning since methane is 80 times more effective in trapping heat than carbon dioxide in its first two decades in the atmosphere.

Methane also accounts for roughly a quarter of all the heat trapped in the atmosphere since the pre-industrial era, according to IPCC data. On the flip side, it also degrades faster than CO_2, making it a primary greenhouse gas target and a crucial part of global decarbonization. Managing methane levels is therefore essential to humanity's efforts to avoid a climate breakdown later in the century.

The assumption in most climate models that carbon release in the Arctic will be an orderly, gradual process may need another look. According to estimates by Arctic scientists, the abrupt permafrost thaw beneath lakes could potentially more than double the permafrost carbon release in the twenty-first century beyond that of gradual thaw alone.

"There is an urgent need for an accelerated scientific effort to more accurately estimate and communicate the likely magnitude of increased carbon dioxide and methane emissions from a warming Arctic," Natali and other researchers wrote in their 2021 paper. "At present, not even the current scientific understanding of future emissions from a warming Arctic is reflected in most climate policy dialogue and planning."

The prospect of mudslides and bubbling methane gas leaks is worrying, given the sheer scale of territory at risk. Alaska is big enough to subsume the combined landmass of the United Kingdom, France, Germany and Italy.

Siberia takes three-fourths of the Russian Federation and is one-third larger than the U.S. and one-fourth more expansive than Canada. The world's carbon-rich permafrost zones in the circumpolar regions also include Iceland, Denmark, Greenland, Finland, Iceland, Norway, and Sweden.

The Arctic plays an outsized role in the global climate future. The permafrost zone in the higher latitudes represents about 24 percent of the Northern Hemisphere's land surface and is the equivalent of a massive carbon storage system, or what scientists call a terrestrial carbon sink.

Inner Mastodon

At the end of the last ice age, massive ice sheets receded to the north, crushing, dragging and freezing tons of organic material from microbes, plants and animals along the way. There's an estimated 1,400 gigatons of carbon frozen in the world's permafrost zones. That's nearly twice the 850 gigatons of carbon scientists believe is in Earth's atmosphere.[7]

One of the more intriguing questions in climate science is how global warming might impact the microbially powered biogeochemical recycling of essential chemical elements like carbon, hydrogen, oxygen, nitrogen and phosphorus in the Arctic.

Soil bacteria convert nitrogen gas from the atmosphere into proteins that allow them and symbiotic plants to grow and thrive. Phytoplankton in our oceans use the sun's energy to convert carbon dioxide in the air and produce oxygen as a byproduct. Other microorganisms consume or produce methane gas and carbon dioxide.

Nobody is quite sure how much potential pent-up greenhouse gas is at risk of being released into the atmosphere if the Big Thaw now underway in the Arctic leads to a microbial feast of newly available organic material as the permafrost unfreezes.

What is clear is that since the start of this century, Arctic temperatures have risen roughly twice as fast as global temperatures and some scientists fear the region is now locked in an accelerated loop of rapid warming, due to a process called the albedo effect, which refers to the ability of surfaces to reflect sunlight.

Historically, the region's ice sheets, glaciers and snow cover reflected the sun's energy back into space. Yet in recent decades, the melting snow and ice have uncovered more land, rocks and water that absorb the heat from the sun.

The unstable Arctic permafrost is sometimes characterized in scientific literature and media reports as a ticking "carbon bomb" on the rooftop of the world. That's probably not the right metaphor. Scientists generally agree there will be no cathartic explosion of emissions from the region in which all the carbon is released at once.

"Not all of the carbon will come out of the Arctic. Maybe it will be 5 percent or 10 percent," reckons Natali. Yet even a net emission of greenhouse gases on that scale in the decades ahead would be unwelcome news.

Bear in mind the razor-thin margin the world has in preventing global warming from exceeding a 2 degrees Celsius increase threshold over pre-industrial levels that experts believe would destabilize human civilization at the end of this century.

Even a small overshoot over that mark might mean a step-change in the intensity of our future climate troubles. "When you think about what that 5 percent or 10 percent means, it's like adding another greenhouse gas-emitting country," Natali said.

There's some evidence that the Arctic permafrost zone is already starting to transition from a carbon sink to a carbon source. In a 2019 study funded by NASA, a team of 70 researchers led by Natali, using field observations, remote sensors and computer models, estimated that the Arctic added slightly more carbon to Earth's atmosphere each winter from 2003 to 2017 than was being absorbed by the region's plants and trees during the summer months.[8]

Their model predicted that CO_2 emissions could increase by 41 percent by 2100 if the planet continues to warm and industrial nations don't take actions to reduce fossil fuel use. If so, this would be a shocking reversal for a region that captured and stored carbon reliably for tens of thousands of years.

Geological Disruption

Climate change skeptics will sometimes point out that the world has survived hotter periods in its 4.5 billion year history. That's true, but really beside the point. What makes the current warming trend so disruptive is the *pace* of change over the last two centuries and the current trajectory of that rate of change going forward.

"This carbon has been captured in the Arctic over this geologically long time-frame of tens of thousands of years," Natali explained. "Now, you're talking about a 100-, 200- or 300-year release time-frame that is actually really short. Geologically, it is something of a bomb. It's not something that we want to ignore."

That said, there is still no clear consensus among climate scientists, ecologists and microbiologists about the future of the Arctic's overall carbon budget, when it might start to unwind or what that might mean for the greenhouse gas dynamics of the atmosphere.

Figuring that out would require a ramped-up, international monitoring effort to better understand the carbon cycling trends from microbes and plants in these colder climes.

Crucially, there needs to be a higher degree of global cooperation, particularly from the Russian Government which controls more than half of the world's permafrost region and whose older scientists aren't always open to sharing geopolitically sensitive data, particularly on greenhouse gas emissions. Getting the geographical coordinates on methane leaks and landslides in Siberia is sometimes tricky for foreign scientists.

The other issue is that most global climate models don't yet fully reflect the unique climate changes in the Arctic, particularly abrupt thaw in the permafrost zones and the massive wildfires that create emissions and burn off vegetation that photosynthetically absorb greenhouse gases.

"Most models don't even have permafrost in them," in Natali's view. "The few that do, depict permafrost thaws as a gradual process," despite the direct evidence that landscape changes are far more non-linear, she pointed out.

Super Models

Global climate models require massive amounts of software code and supercomputer processing power to simulate complex and intrinsically linked systems capable of predicting greenhouse gas emissions decades, even centuries, into the future.[9]

To do so, climate model experts refashion Earth's processes and interactions that flow through the land, ocean and atmosphere into mathematical equations that best reflect that reality.

These models abide by our standard understanding of physics, chemistry and biology. For instance, they reflect the first law of thermodynamics that stipulates energy in a closed system is neither created or lost, but merely changes from one state to another.

So, too, are embedded the mathematical assumptions about how greenhouse emissions shape the Earth's surface temperature and the fluid dynamics of gases in the atmosphere and oceans.

Scientific institutions worldwide such as the United Kingdom's Met Office Hadley Centre and the National Oceanic and Atmospheric Administration's Geophysical Fluid Dynamics Laboratory have developed sophisticated climate models. Another is the Community Earth System Model developed by the National Center for Atmospheric Research (NCAR) in Boulder, Colorado.

If you want a deeper understanding of what the climate models might be missing about the microbial processes shaping planetary carbon cycles, a good starting point is to book a video chat with William Wieder, an Earth systems modeler at NCAR.

Wieder earned his PhD in evolutionary biology and ecology at the University of Colorado in Boulder and still lives in the area. When I caught up with him in the summer of 2021, a miasma of smog from Californian wildfires had recently created a haze in the Rocky Mountains region near him. "Air quality has really been bad in the U.S., particularly in the West," Wieder said. "People are really keen to understand what's going on."

Earth System Models are among the most sophisticated, predictive modeling tools in climate science. They simulate how carbon and nitrogen cycles, together with ocean, land and atmospheric chemistry, respond to increased greenhouse gas emissions and changes in photosynthesizing plants, among other processes.

Wieder is one of a handful of scientists worldwide trying to map and incorporate the behavior of microbes into various soil carbon turnover scenarios that can inform predictive climate models. He has programmed two microbial groups into his model. One whose lifecycle is super-fast and isn't very efficient at using or storing its resources; the other with greater longevity and resource conservation. Yet he concedes that we are a long way off from fully capturing the multifaceted reality of microbes in Earth System Models.

In their quest for the most accurate picture of the Earth's climate system possible, scientists face two big hurdles: the sheer complexity of climate change and the limitations of computing power. The trillions of calculations per second required to model every cubic inch of Earth is an epic challenge that would blow a hole in most research budgets.

To get around this, climate modelers divide up the Earth into a series of large-scale grid cells that represent the atmosphere, oceans and land. The size of these grids typically represents 100 kilometers (62 miles) of the longitude and latitude on the planet. The bigger the grids, the lower the spatial resolution of the model.

Since these are relatively big areas and rough approximations of complex interactions, climate experts fine-tune or "parameterize" their models. They develop computer code to account for regional climate variations or simplify processes that aren't well understood.

Aside from computing and financial constraints, there are scientific discovery issues as well. There are still gaps in our knowledge of the size and scope of the microbial world and the precise role they play in biogeochemical recycling of carbon and nitrogen. There's no easy way to distill the diversity of soil microorganisms to a functional classification that you can apply worldwide and reflect in a nuanced and predictive way in the climate models.

"Take a big spoonful of soil from your backyard, there are as many microbes there as people on the planet in raw terms," marveled Wieder. "The diversity of these microorganisms is pretty astounding. We still don't really know what they're doing."

As a result, the "uncertainty about potential Arctic greenhouse gases is huge," according to Wieder. A lot of models assume gradual permafrost thaw and the Arctic continuing to play historical roles as an absorber, rather than as an emitter, of greenhouse gases until the end of this century. "I don't think that's realistic," he said.

Many of the big-impact thawing events at higher latitudes are sudden, or nonlinear, and on a spatial scale too small to be reflected in a climate model's grids. Take, for instance, the phenomenon of ice wedge melts that occur under bodies of water.

"When that ice wedge melts, it's like pulling the drain out of your bathtub and all of the water above it drains out. We don't get those dynamics," explained Wieder. "When the water is gone, the soil can warm up a lot faster. So the ice wedges are important."

Another data gap in the climate models, according to Wieder, are the world's peatlands. These refer to the moss-covered lands and marshes that span from the forested bogs in Europe and highlands of the Andes in South America to Southeast Asian tropical peat swamps. They collectively store more gigatons of carbon than the world's forests and are among the world's largest natural terrestrial sinks.

Damaged wetland systems from ocean pollution and coastal economic development are releasing nearly 6 percent of humanity-driven CO_2 emissions. "That's a big deal," said Wieder. "We don't have moss in the model."

Moving Targets

If greenhouse gas emissions from permafrost or peatland emissions are overshooting the base-line assumptions in our climate models, we may be considerably further behind in our race to hit internationally agreed global warming targets.

The Paris Agreement, adopted in 2015, aimed to keep global temperatures from rising no more than 2 degrees Celsius, preferably below 1.5 degrees Celsius, over pre-industrial levels.[10] Greenhouse emissions have kept trending upward, with the exception of 2020 as the pandemic brought the global economy to a standstill.

In November, 2021, many major economies participating in the United Nations Climate Change Conference, known as COP26, agreed to net-zero carbon targets by 2050 and an accelerated phasing out of government subsidies for coal, oil and natural gas. There were a lot of promises made by governments and the private sector, but skeptics were disappointed by a dearth of hard targets and deadlines to meaningfully arrest the growth in fossil fuel emissions.

The aspiration of containing the rise in global warming within 1.5 degrees Celsius still lives on. Less optimistic scientists believe that's unlikely, and even avoiding a 2 degrees Celsius increase in global warming over pre-industrial levels will be a tall order. If our baseline assumptions of emissions growth from rapid thawing and microbial greenhouse emissions are way off, more extreme measures might be needed.

It's conceivable that humanity's insurance policy against climate breakdown will require two incredibly daunting tasks: Decarbonizing the global economy while simultaneously launching potentially risky geoengineering projects to physically remove greenhouse gases from the atmosphere.

Climate change is not anything like the orderly, linear mathematical equations reflected in current climate models. The pace of the planet's transformation races ahead of our grasp of the long-term implications. That's dangerous because once these climate changes hit a phase transition point, they're nearly impossible to unwind.

"Nobody ever expected ice sheets to be melting as fast as they are. Nobody expected fires at 1 degree Celsius to be as bad as they are," according to Arctic ecologist Natali. "If you told me 10 years ago, there would be exploding (methane gas) craters, I would have said you are crazy. We don't know all of the processes, because we are going into a space that humans haven't been in."

We are already seeing evidence of some strange evolutionary adaptations within the Arctic's microbial realm. The ice melt from the permafrost thaw has swelled the levels of Alaskan rivers and discharged huge amounts of organic material and terrestrial microbes into the Arctic Ocean.

The frozen soils in the northern latitudes are home to both methanogenic archaea that consume carbon and release methane, as well as methanotrophic bacteria that feast on methane gas for energy. As the permafrost zones thaw and climate warms, a big population shift in favor of one or the other microbial tribe may have big implications for the climate later in the century.

One metagenomic study found evidence that land-based *Actinobacteria* species had transferred genes for processing carbon to aquatic planktonic *Chloroflexi* bacteria.[11] These types of horizontal gene transfers among microbial actors in permafrost regions could change carbon cycling dynamics in unpredictable ways and our scientific understanding is still quite limited.

Degrees of Concern

Since the last ice age ended more than 11,000 years ago, a period that scientists formally call the Holocene Epoch, the Earth's carbon cycle and climate system have been synchronized in an advantageous way for the flourishing of human civilization.

Then starting in the late eighteenth century came the industrialization of the West, and with it the emergence of capitalist systems that came to rely on the extraction of natural resources and the burning of carbon-based coal and oil to power their economies.

Since then, the expanding miasma of greenhouse gases hovering in the atmosphere has started to overwhelm the planet's ability to absorb it. Every bit of additional warming is a degree of concern. The difference between a 1.5 degrees Celsius vs. a 2 degrees Celsius temperature increase by the end of this century would be pronounced in terms of destructive impact.

Heat waves, storms and sea level rises would be about one-third, or 33 percent, more intense, according to a study by the European Geosciences Union.[12] Tropical corals would be wiped out, while water shortages would turn critical in some regions.

In what UN Secretary-General António Guterres has called a "code red" alarm for humanity, the latest IPCC climate assessment released in August of 2021 and based on the research of 234 scientists from 66 countries makes for some unpleasant reading.[13] As of 2019, atmospheric concentrations of carbon dioxide, methane and nitrous oxide are at levels not seen in historical measurements for roughly 800,000 years.[14]

Climate scientists are now unanimous in their warning that without emergency measures to eliminate greenhouse gas pollution, the planet would blow through an internationally agreed warming target of 1.5 degrees Celsius over pre-industrial levels in the next two decades, the U.N. assessment warned.

Amazonian Carbon Shift

If the planet can be said to have a respiratory system, the Amazon basin might be its lungs. With its forests and luxuriant plant life stretching from the Atlantic to the Andean foothills, the region is the world's most diverse biological reservoir, home to millions of species of plants, birds, and life forms still undiscovered by science.

The swath of tropical jungle spans nine countries, accounts for approximately 40 percent of Brazil's land mass and is twice as big as the rainforest regions in Africa's Congo basin and Indonesia.

The Amazon region's microbial systems and botanical life are greenhouse gas recycling and storage systems, locking up massive amounts of carbon in tree trunks and soil. However, a 2015 study estimated the region's uptake of carbon had fallen by one-third compared with the 1990s.[15]

Decades of relentless burning, logging, mining and development have degraded the region's carbon-storage ability. Along the southern and southeastern edges of the rainforest controlled by Brazil, there's been an arc of deforestation to make room for soybean crops and meat production. In Peru, the palm oil industry and artisanal gold mining are among the key drivers of rainforest destruction.

In Colombia, ironically, the 2016 peace accord between the central government and the revolutionary paramilitary group called FARC previously based in the Amazon has opened up more of the basin to cattle grazing.

While the rate of deforestation in the Amazon has moderated since 2004 due to new protected areas, satellite monitoring and global environmental pressure, the cumulative damage may have changed the greenhouse gas balance in a detrimental way.

In March, 2021, the world received a bracing wakeup call when a comprehensive study of the region looking at a variety of natural and humanity-generated gases concluded that parts of the Amazon Basin may now be a net emitter of greenhouse gases.[16]

Carbon dioxide is just one variable in the multi-faceted climate change calculus. Perhaps even more worrisome, the region is now emitting large volumes of methane and nitrous oxide, both of which pack more global warming power than CO_2, due to increased cattle production and the drying out of seasonally flooded forests.

A big source of greenhouse gas emissions from Brazil comes from land use change, particularly the conversion of rainforest to cattle pasture. Brazil is a leading meat exporter, especially to China. Tearing down forests for cattle has massive implications for the microbial balance in the Amazon.

In 2020, a team of researchers led by microbiologists Klaus Nüsslein and Marie Kroeger at the University of Massachusetts Amherst sampled methane-cycling microbial systems in the Brazilian states of Pará, home to the Tapajós National Forest, and Rondônia.[17]

What they found was a major expansion in the population of methane-producing microorganisms called methanogens. Cattle and other grazing animals have evolved to host these microbes that help break down hard-to-digest grass, and produce methane as a byproduct.

Puffs of methane are emitted by livestock as they belch and through their manure, where poop-eating methanogens produce even more of that gas. The global demand for more animal protein has led to a massive expansion in the number of cattle, swine and chickens domesticated for food production. Grazing animals are estimated by the United Nations to account for 14.5 percent of annual methane emissions, accounting for more than any other single source of the greenhouse gas.

Oceanic Roulette

Equally as worrisome, big changes may be taking place among microbial networks in our oceans. There's some urgent discovery science that lies ahead, starting with the future behavior, diversity and population growth of marine phytoplankton that are responsible for half of the world's oxygen production and CO_2 recycling.

Two main classes of phytoplankton, known as dinoflagellates and diatoms, devour carbon dioxide on the same scale as the world's land plants and forests. Some is recycled to other creatures that feed on these microbes. Other CO_2 descends to the deep ocean when phytoplankton die.

While trees and plants are locked in place and evolve over centuries, marine microbes are mobile, widely dispersed around the globe and have life spans and mutation rates numbering in days, not years. That means that their adaptive changes, for good or for ill, to climate shifts could be relatively quick.

When microbes face existential risks, they often seek out more favorable environments. In a pinch, they can also swap genes and change their molecular mechanisms and biogeochemical cycles in ways that scientists don't fully understand yet.

Marine microbes aren't hard to find in the oceans. They represent a vast landscape of life: bacteria, archaea, protists, fungi and algae, not to mention associated viruses like bacteriophages. There are a billion-plus microbes in each liter of seawater.

As multitudinous as they are, these microbial swarms have traditionally been difficult to culture in labs and study up close. Researchers are still trying to understand all the symbiotic or competitive relationships within microbial communities, how all of their genes work or whether climate growth is depressing or enhancing their populations.

Our visibility into the inner workings of these microbial colonies is improving somewhat with the arrival of metagenomics, in which large DNA samples taken from the wild can be sorted, sequenced and read for clues about biodiversity. Spatially mapping populations of bacteria can reveal new environmental pressures and the health of an ecosystem.

The first oceanic big census of marine microbial life called the International Census of Marine Microbes (ICoMM) ran from 2004 to 2010 and there have been various research projects like the Global Ocean Sampling Expedition and Tara Oceans.

More research is needed to understand the distribution of microbial genes, how microbial communities come together and disband, as well as their distribution and composition across time.

Will the increasing acidification of our oceans and lakes decrease their numbers and leave vastly more heat-trapping carbon dioxide up in the atmosphere? Might warmer and more acidic ocean water tilt the competitive advantage toward viruses that attack photosynthesizing bacteria that recycle carbon? One study forecast that if viral phages were able to reduce the photosynthetic power of cyanobacteria, that process could add up to five billion metric tons of CO_2 into the air annually.[18]

Deus Ex Machina

The growing sense of urgency to reduce planet-warming carbon emissions has led to calls for deploying radical geoengineering technologies that can buy us more time. Once viewed as the province of science fiction novels and too risky to contemplate, these climate system interventions are now getting a serious look by major universities in the U.S., United Kingdom and China.

Former British government chief scientist Sir David King, a climate change expert who established the Centre for Climate Repair at the University of Cambridge, thinks we'll need to launch ambitious geoengineering projects to make the emissions math work and allow humanity to work through the current climate crisis.

Large-scale geoengineering projects to extract greenhouse gases are backed by Big Oil companies like Exxon-Mobil, Chevron as well as the Bill & Melinda Gates Foundation. They are viewed with great suspicion by some environmental groups. Critics dismiss investments in carbon-sequestering as empty "techno-solutionism," whose unspoken agenda is to protect the corporate power and profits of the fossil-fuel industry.

Cambridge University's King doesn't see carbon-extraction strategies as an artful dodge from decarbonizing the global economy. He argues that we urgently need to get to net-zero emissions of new greenhouse gases by mid-century.

In his view, that will require transitioning away fossil fuels to renewable and nuclear energy sources, by decreasing demand for beef, soya and palm oil that power agricultural greenhouse emissions, and by expanding reforestation projects, among other measures.

Even if we manage that, however, it still won't be enough given the climate damage already endured by the planet. King claims that we will need to expand existing carbon sinks in our oceans and on our lands to lower greenhouse emissions that are on track to surpass 500 parts per million (PPM) to a more sustainable level of 350 ppm by the end of this century.

Finally, there's what he calls "climate repair" projects. One technology that's worth serious consideration, King contends, is ocean surface iron fertilization. The idea is inspired by a natural process. As wind blows over the Sahara, iron-enriched sand particles and deposits are swept up and transported over the Atlantic.

Scientists have known for decades that microbial phytoplankton are attracted to iron, which is a crucial nutrient missing in most oceanic food webs. In ocean waters with the right conditions, an infusion of iron particles triggers the formation of plankton blooms that also serve as a biological pump to absorb heat-trapping carbon dioxide from the Earth's atmosphere.

One of the early proponents of the "iron hypothesis," biogeochemist John Martin, half-jokingly quipped at a late-1980s seminar at the Woods Hole Oceanographic Institution: "Give me a half tanker of iron, and I'll give you an ice age."

Researchers have observed that regions of the Atlantic that receive pepperings of Sahara dust enjoy rapid greening, as microbial phytoplankton in search of iron migrate to these areas. There's also a big uptake in CO_2 absorption.

More research is needed to understand the possible unintended consequences of this climate system intervention. Would it make sense to intentionally introduce more iron into our seas and how would that work?

There's the risk of overfed phytoplankton creating huge blooms with toxins that harm marine life. Others wonder about the ecological impact of iron filing deposits in our oceans.

"If we were able to green 3 percent to 4 percent of the ocean's surface every year by dusting it with iron, we could achieve 30 billion in greenhouse gas removal," King said in a public lecture. "That's attractive technology to look at, but there are potential downsides."[19]

The most audacious geoengineering projects being contemplated focus on the world's permafrost zones near and within the Arctic Circle, which is warming at a much faster rate than the rest of the world.

"The profound danger signaled by high Arctic temperatures is that a tipping point has passed," King noted in a speech. The accelerated warming and ice melt in the Arctic risk spilling over to Greenland, the second largest ice body in the world that holds enough water to raise global sea levels by 7.5 meters, or 23 feet, according to King.

Technologies being explored by the Cambridge climate repair centre include refreezing rapidly warming sections of the polar regions by injecting salt particles from the ocean into the clouds. The aim would be to make them brighter and reflect more solar energy away from the lower atmosphere.

Harvard researchers, financially backed by Microsoft co-founder Bill Gates, are exploring solar geoengineering to reflect sunlight away from Earth by spraying non-toxic calcium carbonate dust into the atmosphere.

It's a radical approach plenty of other scientists are wary of, given the potential unknown impact on complex weather patterns. Environmentalists worry that such grand schemes will create false hopes and disincentivize political leaders from pushing hard to phase out fossil fuel energy.

The climate change crisis has been centuries in the making and finding our way out will take several generations of effort as we transition to new ways of organizing our economic life and powering our societies.

Such structural changes require interlocking and self-enforcing trends such as technological innovation, global shifts in capital, new energy sources and changes in public and political opinion. Biotechnology, microbial science and new specialities like synthetic biology could chart a pathway toward a more sustainable future. Advances in life sciences, computer automation and artificial intelligence are remaking industries from health and agriculture to energy and materials.

The hype coming from the management consultants and corporate marketers about the coming Age of Biology should be greeted with a healthy dose of skepticism, as with any other proclamations of new economies and eras. Yet there's no denying that the speed with which researchers can sequence, synthesize and redesign microbes for medical research and commercial applications is potentially transformative.

Twentieth-century capitalism created a lot of wealth, lifting millions out of poverty, with a linear economic model: extract, produce, consume, and throw away. That take-make-waste approach has locked us into unsustainable patterns that are destabilizing the planet. Microbial and bio-based materials that can be recycled from natural systems offer a way to design more circularity (reuse, remanufacturing and recycling) into economic life that causes less damage to the environment.

The coming biology revolution also carries big risks: The cadence of technological change will challenge our ability to manage it responsibly. Biology is a powerful and self-replicating phenomenon that can cause lasting damage in the wrong hands. Such are the paradoxes that await humanity in the decades ahead that we shall turn to next.

THE DEEP FUTURE

Chapter 12

The Synthetic Biology Paradox

In early 2020, just as the world started to grasp the gravity of a novel coronavirus rampaging through Wuhan, China, a team led by synthetic biologist James Collins at the Massachusetts Institute of Technology published a remarkable finding in a scientific journal called *Cell*.[1]

The group used an artificial intelligence (AI) tool known as machine learning to discover a new class of antibiotic drug that's effective against dangerous strains of the bacterium *Escherichia coli* and drug-resistant tuberculosis.

Collins' team designed a program called a deep neural network — algorithms that mimic the neurons in our brains and do sophisticated computations on massive amounts of information — that screened biological databases, searching for molecules that might be effective against *E. Coli* and weren't already used in conventional antibiotics.

The AI bots found roughly 100 potential candidates for lab work, and one of them turned out to be a structurally unique antibiotic. Collins' team called it halicin, a homage to HAL, the menacing starship computer of *2001: A Space Odyssey* fame.

Collins has described halicin as "arguably one of the more powerful antibiotics that has been discovered." In tests, the newly discovered antibiotic was effective in killing hard-to-treat bacteria such as *Clostridium difficile*, *Acinetobacter baumannii*, and *Mycobacterium tuberculosis*. The only bacterial species it couldn't slay was *Pseudomonas aeruginosa*, a lung pathogen. That said, halicin hasn't yet faced a regulatory barrage of clinical trials to validate its safety and effectiveness.

Former Google Chief Executive Officer and Executive Chairman turned venture capitalist and synthetic biology investor Eric Schmidt sees the union of AI and biology as potentially transformative. Innovation Endeavors, a fund back by Schmidt, has set up an incubator project called Deep Life to identify startups that can apply computer science — specifically AI applications such as machine learning — to accelerate the discovery of new therapeutics, diagnostics and industrial life science.

"Math enables you to understand physics and AI will allow us to understand biology," Schmidt told me in an interview from New York. The remarkable thing about the MIT breakthrough, said Schmidt, is that "they took 100 million chemical compounds, embedded it into an AI system, and it figured out how the proteins and molecular chemistry would work without them telling it. The program literally discovered it."

The former Google executive thinks U.S. and European pharmaceutical companies may be on the cusp of a brilliant run of drug discovery as AI unlocks new features of microbes and chemical combinations that humans may miss.

"We need to build big biobanks (databases of biological data) with this kind of information that accredited researchers can build on top of that, along with other various bioengineering tools to accelerate all of this," Schmidt told me, adding that China is also chasing AI-powered drug development.

In a National Security Commission Report on AI released in early 2021 that Schmidt oversaw, synthetic biology was identified as a strategic life science specialty the U.S. and China would hotly contest in the decades ahead.[2] The country that masters AI first would have a decisive advantage. The U.S. is in the lead for now, but China is coming on strong, the report concluded.

Synthetic biology is still such a new field that there's not yet a commonly agreed-upon description of what it does. One way to think about it might be this: It's a scientific endeavor that uses engineering methods to discover, design and build biological systems that improve life.

Since the mid-1990s, an eclectic group of computer scientists and engineers, and MIT's Collins is certainly one of them, have pioneered tools and procedures to find new ways of exploring microbes, enzymes, genetic circuits and other biological systems.[3]

Despite its futuristic and somewhat unsettling name, synthetic biology draws its intellectual influences from well-established disciplines —

microbiology, computer and software engineering, robotics, genetics, systems biology, and biophysics, among others.

What's distinctive is SynBio's preoccupation with finding practical and effective ways to decode the chaotic world of biological cells. Biological research has made a momentous shift in recent decades as the tools to decode and design cells have grown in sophistication. Researchers have figured out better ways to understand the physical details of microbes, genes, protein properties and the interaction of biochemicals.

When synthetic biology came on the scene in the late 1990s and early 2000s, its practitioners imported valuable lessons about standardized, modularized and replicable engineering from the realms of semiconductor manufacturing, software engineering and computer science.

In what's now an enshrined piece of Silicon Valley lore, Intel Corporation cofounder Gordon Moore famously predicted in 1965 that computing would increase exponentially in power while decreasing in cost.[4] Moore's Law correctly foresaw a compounding dynamic in semiconductor technology that eventually opened the path for society-altering innovations such as personal computers, the Internet, wireless technology, smartphones, and artificial intelligence.

SynBio true believers think that programmable cells are in a way the new silicon. The future belongs to companies and economies that can harness the vast genetic information and biochemical processes stored in the DNA of microbes and other living organisms. Those gains can be locked in and serve as the foundation for accelerated progress going forward, much as a next-generation microprocessor with enhanced computing power spawns new applications.

Optimists envision the sunny uplands of a twenty-first-century bioeconomy, featuring more effective medicines, healthier foods, and sustainably produced consumer products, materials and energy. Synthetic biology skeptics see a technology with huge risks, perhaps even existential ones, that we will explore in greater depth in another chapter.

For the moment, it's worth considering that the microbial world is saturated with information that we've only begun to fathom. Microbes have been evolving and designing ways to metabolize energy, fend off predators and thrive in challenging environments for nearly four billion years. Along the way, these microorganisms have developed biochemical talents and molecular pathways that, if fully understood, may be very beneficial to humanity.

Consider the genomic information from a microbe, mammal or human that fits compactly in the DNA of a single cell. It's an information storage unit of incredible density, many orders of magnitude beyond any physical database or server farm network. Run a random teaspoon of dirt or seawater through a sequencer, and you'll find gigabytes of genetic information.

The coding language of DNA isn't the digital bits and bytes of computers and other digital devices, but the series of As, Cs, Gs and Ts that represent the quartet of corresponding building block nucleotides called adenine, cytosine, guanine and thymine. The variations in sequences of these four information-carrying molecules are the source code of biological life. Scientists use high-powered genetic sequencers to read it and DNA synthesizers to write it.

As young a discipline as it is, synthetic biology has already delivered some meaningful contributions to human health. In the infectious disease world, it's credited with discovering a way to manufacture artemisinin, a key ingredient in anti-malarial drugs that's derived from the Asian sweet wormwood *Artemisia annua*. Discovered by Chinese scientists in the 1970s, global demand sparked shortages and price spikes of the drug compound in the early 2000s.

A team led by Jay Keasling, a chemical engineering and bioengineering professor at the University of California, Berkeley, came up with a way to genetically modify yeast to make artemisinin.[5] In 2014, French drugmaker Sanofi started distributing the drug just as African nations were coping with malaria outbreaks. Synthetic biology techniques were crucial in the speedy rollout of COVID-19 vaccines as well.

SynBio Godfather

If the rapidly evolving field of synthetic biology can be said to have an intellectual muse, it might be Tom Knight, the long-time, former MIT computer scientist and synthetic biologist entrepreneur.

During his multidecade teaching and research run at MIT, Knight embodied the university's geeky culture of brilliant overachievers. A Boston-area native, Knight took MIT courses in computer science and organic chemistry while still in high school and spent summers in his teens working at the university's artificial intelligence lab.

In the 1960s and 1970s, as a full-time MIT student and later as a faculty member, Knight co-engineered the Advanced Research Projects Agency Network , or ARPANET, an early version of the Internet. He built operating systems and designed computer workstations. Knight eventually earned more than 30 patents for his discoveries in computer science and electrical engineering. He also worked on one of the first silicon retinas, a chip that mimics the neural circuitry of the human visual system.

Around 1990, Knight said he started to realize that Moore's Law might eventually hit a physical limit at some point. As the miniaturization of transistors — electrical gates that switch on and off to perform calculations — hit the 10 nanometer range and below, the engineering challenges become considerable. That said, there remains a lively debate in Silicon Valley in the early 2020s over whether Moore's Law is finished just yet.

In any event, the 1990s version of Tom Knight made a fateful decision to shift his career from bits and bytes to cells. Then in his fifties, Knight enrolled in undergraduate biology courses and by 1997 had secured funding from America's Defense Advanced Research Projects Agency (DARPA) to open a biology lab in MIT's computer science department. In those days, researchers interested in the fusion of engineering and biology called their nascent field cellular computing.

Early on, Knight remembers talking to biologists at MIT and elsewhere and coming away surprised with their sense of certainty that many of the basic research questions in microbiology had already been figured out. "Tom, why are you working in *E. Coli*," Knight recalls one prominent biologist asking. "We already understand everything there is to know about that organism."

Knight respects biologists, but isn't so sure it's time for scientists to take a victory lap about cracking the code on the genetic makeup of individual microbes. He learned that lesson early on in his newfound biology career, when he started working with a class of bacteria called *mycoplasma*. In those days, Knight used a standard research technique called "transposon-generated knockouts," in which a researcher inserts pieces of DNA into the genome of an organism with the aim of a displacing, or knocking out, a specific gene.

If you remove a gene, and the organism dies, a biologist knows that it serves an essential function. "I discovered 400 or 500 genes that were essential," Knight explained. However, there were 5 percent to 10 percent of them that are crucial to bacterial life, and common across a lot of organisms, that biologists remain completely baffled about, according to Knight.

"You see these genes widely across the biological spectrum, and there's no known function for these genes," he said. "You take them out. The organisms die. That's the situation we are in right now. Basically, no progress on what genes do."

While there's plenty more basic research ahead in synthetic biology, there's already a lot of useful applied work underway. Knight and other engineering-minded biologists have imported a different perspective to the field. "I'm an engineer. A lot of my motivation," he said, "is can I twist whatever is out there and make it do something useful."

In the early 2000s, Knight and other scientists found a more efficient way to engineer microbes and cells with an innovation called BioBricks.[6] They created a database of DNA sequences that were standardized, tagged and stored for future cell design projects. The aim was to save researchers work and time by archiving modular, biological building blocks of proteins and enzymes that could be assembled and reassembled for prototypes of new drugs and products.

The first worldwide gathering for the new discipline, Synthetic Biology 1.0, took place in the summer of 2004 at MIT. Around the same time, Knight and his then-MIT colleagues Drew Endy, Randy Rettberg and Gerald Sussman organized iGem, or the International Genetically Engineered Machine competition.

University students working in multidisciplinary teams explored the frontiers of synthetic biology by building new products and features with genetically altering microbes, proteins and enzymes. Knight and others also created the somewhat ghoulish sounding Registry of Standard Biological Parts, a repository where synthetic biologists could digitally catalogue and physically store genetic sequences and parts for future research projects.

Since BioBricks, other SynBio assembly methods such as Golden Gate and Gibson have gained widespread use in the field. Researchers have also developed a common computer programming language called Synthetic Biology Open Language so that software tools can understand and manipulate descriptions of biological parts and genetic processes.

In 2008, Knight left academia for Ginkgo Bioworks, a biotech firm in Boston that calls itself an organism company. It was co-founded by Knight and four MIT PhD students (Austin Che, Reshma Shetty, Jason Kelly and Barry Canton) in bioengineering and computer science. Knight helped line up early government funding and invested some of his own money into the startup.

These days Knight sports a snow-white, chinstrap beard that gives him a slightly Victorian vibe. Yet his mindset is very much pointed toward the deep future, toward unraveling more of the secrets of microbes and the genes that shape them.

Engineering new proteins for pharmaceuticals and consumer products involves a lot of interesting technology, but it is at heart quite a straightforward process, according to Knight. "We engineer good proteins by looking at another protein that almost does the job that we want, or does it poorly," explained Knight. "Then we go and we take that protein sequence and we look at all of the other 100,000 proteins that are in the databases."

Thousands of other proteins are analyzed, sampled and sequenced for comparative analysis. Then a new protein prototype is designed and built using DNA synthesizers. After a lot of refinements, "you almost always discover that you have ones that are 2X, 5X, 10X, or even 20X times better than the ones you started with," said Knight. "That's the dirty little secret about how we engineer good proteins."

Ginkgo has teamed up with Fortune 500 companies and new ventures to develop strains of microbes to make fertilizers, fragrances, and break down pollution. During the COVID-19 pandemic, it helped the nearby Cambridge-based firm, Moderna Inc., scale up production of its successful mRNA vaccine. Ginkgo, which went public in 2021, trades under the stock ticker DNA.

A number of synthetic biology and biotech firms such as Twist Bioscience, Agilent Technologies, and Zymergen have developed DNA sequencing and cell manufacturing platforms for in-house work or that they outsource to companies and joint venture partners.

These SynBio players offer pharmaceutical, food and consumer products companies a way to enhance their own research and development efforts. Some companies prefer to contract out part of the sophisticated work it takes to tease out new drug therapies, food ingredients and biomaterials from microbes, proteins and enzymes.

Ginkgo has developed biological foundries, in which microbes and cells are designed, prototyped, tested and refined in an automated manufacturing system that employs robotics, data analytics and software.

A typical project starts with Ginkgo engineers working with clients on computational design tools to virtually assemble the desired DNA or microbial trait. The sequences are then sent electronically to a DNA printer, which uses something similar to ink-jet technology to dispense tiny droplets of base chemicals to create genetic segments. Once that's done, the newly designed genetic material is inserted into cells and treated with chemicals and heat.

Initially, the newly created cells are often just a smudge of chemicals in a tiny tube. The next step is to scale up and grow billions of copies, usually by using small bioreactors that typically rely on a fermentation system similar to what one might find in a brewery.

The batch of cells is then sequenced again to make sure the design specification hit the mark. The cells are analyzed with high throughput screening equipment that can test thousands of samples for biology activity at the cellular and biochemical levels.

Throughout the entire process, Ginkgo software systems capture performance and metadata (essentially, data about other data) that can be analyzed and repurposed for future work. The company has processed approximately 400 million proprietary genetic sequences and has genetically designed DNA samples that are sitting in a freezer ready to be utilized quickly to launch new projects.

Looking into the deep future, Knight is hesitant to make big predictions about where synthetic biology will take the world, other than to say that in a decade we won't recognize the technology. "I don't think we will be reinventing dinosaurs," laughs Knight, but there will be examples of really sophisticated biology such as human organ transplants from pigs that may surprise people. "We are going to start building new things. That's going to be a very powerful tool in helping us figure out what's really going on in biology."

SynBio-phoria

Silicon Valley and Wall Street see synthetic biology as the most powerful technology of this century, with vast commercial potential. Silicon Valley royalty who made their fortunes in the digital age such as Bill Gates (Microsoft), Schmidt (Google), Marc Andreessen (Netscape) and Jerry Yang (Yahoo!) are all SynBio investors.

The long-term potential of synthetic biology is often viewed through the prisms of electrical and software engineering. DeepMind research scientist and computational biologist Sara-Jane Dunn likens microbial and biological cells to mini-computers. Our cells run on genetic software programs that use inputs and outputs to interact dynamically with their environments.

These programs work in a distributed way within our bodies. If we can better understand these living software programs, we can develop new disease-fighting strategies. "If we truly understood these biological programs, we could debug them when things go wrong," as Dunn has explained it.[7]

True enough, but actually manufacturing new biological entities is a physical process involving chemicals. It's an enormously complex one that's tricky to pull off at industrial scale.

In 2021, the stock of newly listed biotech darling Zymergen plunged and its CEO resigned after manufacturing challenges with a key product called Hyaline, an optical film made from a biomolecule that electronics companies could potentially use as an alternative to traditional chemical materials.

Gingko has generated a lot of buzz and its most enthusiastic investors believe it might become the Amazon or Microsoft of biology. Yet so far, the synthetic biology phenom is a company with modest scientific achievements and its labs haven't yet generated a run of smash-hit commercial products that have transformed a company, industry or the broader U.S. economy.

Perhaps that day will come for SynBio. The horizon scanners at McKinsey Consulting see an emerging biorevolution that has the potential to be far bigger in scale than the global impact of digital technology. Some 60 percent of the non-biological or physical inputs of the economy — plastics, fuels, clothing, construction materials, transportation, and so on — could in theory be modified or produced biologically, according to the consultancy.[8]

Like any fast-growing industry with commercial potential, synthetic biology will surely have its share of carnival barkers and rosy forecasts that just don't pan out. Yet there's no denying that industrial biotech is currently the beneficiary of a self-reinforcing cycle of biological and microbial research breakthroughs. The rise of affordable and effective genetic sequencing and synthesizing tools have accelerated innovation in the emerging life sciences field.

It's easy to forget how far we've come from the start of this century. In 2003, it cost about $3 billion for researchers to map out the first draft of the human genome. Now, it costs about $1,000. In addition, ordinary citizens can send saliva samples to 23andMe and have their DNA genetically analyzed on a genotyping chip (a different process than sequencing) that flags health risks and tracks one's ancestry for several hundred dollars. The cost of DNA sequencing has been decreasing at a far faster rate of change than the declining semiconductor prices predicted by Moore's Law.

The COVID-19 pandemic sensitized the general public, or at least the better-informed segments of it, to the big changes stirring in the biological world. Synthetic biology and genetic editing played a crucial role in the messenger RNA and virus vector vaccine technologies that delivered effective drugs in months, rather than years.

Synthetic biologists are now locking in the gains from a decade of advances in computing, data analytics, machine learning, artificial intelligence, and biological engineering. AI applications can sift through gigabytes of genetic information, while the arrival of the CRISPR editing system allows researchers to rewrite the code of life with greater precision.

If synthetic biology enthusiasts are correct, over the next few decades we may see cancer treatments, in which microbes such as bacteria and phage viruses target tumor cells more effectively than radiation and chemotherapy.

Immunology-boosting strategies are also emerging. One is CAR-T therapy, in which immune system T-cells taken from a patient are genetically modified and then put back into his or her bloodstream with enhanced ability to find and kill cancer cells.

Researchers at Stanford are trying to boost the effectiveness of the treatment, which in 2021 only worked in about one-third of clinical trial patients, by looking at what role human microbiota in the gut play with the immune system and cancers.[9]

By the 2030s and 2040s, the human diet may look very different as well. Yes, we'll likely still be eating meat from livestock slaughtered in factory farms. Yet bipedal carnivores concerned about the agricultural industry's carbon footprint will have other options, including synthetically produced meat produced in a lab.

Plant-based steaks, sausage and seafood may taste closer to the real thing as specialty SynBio firms exploit microbial enzymes and other biochemicals to improve the taste and texture of their offerings. These alternative meat products could allow humanity to dramatically downsize a livestock industry that places huge environmental demands on the planet. Biopolymers that safely degrade may start to displace petrochemical-based plastics, while other materials such as synthetic silk and alternative dyes will reduce the carbon impact of the fashion and apparel industries.[10]

A big challenge for human implants and neuroprosthetics has been bacterial infection. New materials that secrete antibiotics over long periods of time are starting to address that problem, according to synthetic biology optimists. In the distant future, imagine nanoscale biosensors inside of us monitoring our vital signs for trouble or alerting us to cancer triggering mutations in our cells.

The boundaries of human intelligence and life spans may well be expanded through human-machine interfaces that boost our ability to store and process information. In the future, our DNA might be repurposed as genetic circuits that we can turn on and off, or even as a deep data storage system.

Researchers have already successfully engineered DNA so its As, Cs, Gs and Ts correspond to 0s and 1s of digitized information. In 2013, scientist Nick Goldman and a team of researchers at the European Bioinformatics Institute designed a DNA sequence that encoded all of Shakespeare's sonnets and an audio file excerpt of Martin Luther King Jr.'s "I have a dream" speech.[11]

They sent the sequence to Agilent, a life science company spun off from Hewlett Packard in the late 1990s, which converted that information into a smudge of DNA at the bottom of a test tube. It was reversed engineered back to a digital file and played back with perfect accuracy.

Some experts believe that at some point humanity's datasphere, all of the bytes created, copied, and consumed annually, will outstrip our ability to physically store it. The world was on track to generate 74 zettabytes (or 74 trillion gigabytes) of data in 2021.[12]

Managing that data on servers, computer hard drives and magnetic tapes is energy-intensive and not great for the environment. Also, the equipment needs to be replaced and upgraded after a decade or the quality of the data is compromised.

What if we could repurpose DNA as a long-term data archive? Doing this practically is decades off, but in theory you could store all of humanity's digital output in DNA strands that would collectively fit into the cargo area of an SUV. And if you stored the DNA in a cold, dark and dry space, the data integrity would last hundreds of thousands of years.

Massive Knowledge Gaps

As impressive as these gains have been, there are gaps in our knowledge of the molecular mechanisms of microorganisms and the intricacies of cellular behavior. One immediate challenge is the sheer vastness of the still-uncharted genetic information in the biological world.

Critics, conversely, find it alarming that, with little public debate or consideration, a fast-growing industry has amassed the power to design living organisms from scratch, whether that's by tweaking nature's molecular machinery or inserting an artificial genome created by DNA synthesis technology into cells.

Today, a clever high school student can use computerized genetic-design tools to program cells and then pay an online DNA synthesizing firm a couple of hundred of dollars to build it out and mail it back.

Commercial kits to perform CRISPR gene editing can be procured on the Internet. There's an international community of biohackers using interchangeable biological parts and standardized techniques to create, build and test their new designs for fragrances and medical treatments.

During a 2018 biotech conference in San Francisco, a biohacker named Josiah Zayner, who runs a genetic engineering company called Odin, used the CRISPR gene-editing tool to inject himself with a therapy to improve the muscles in his forearm.

"This is the first time in history that we are no longer slaves to our genetics," Zayner told a journalist. "We no longer have to live with the genetics we had when we were born. Technologies like CRISPR and other genetic modification technologies allow adult humans to modify the cells in their body."[13]

Is this really a game for amateurs? Biological life is self-replicating and shape-shifting. A genetically engineered change may have unintended consequences that will be felt across generations and entire ecosystems.

Lessons from past scientific mishaps are worth considering. In 1956, geneticist Warwick Kerr cross-bred European and African honey bees imported into Brazil in an effort to improve honey production. It worked; however, a year later colonies of these bees escaped from Kerr's lab.[14]

Thus began the reign of Africanized "killer" bees in the Americas. When provoked, they attack in greater numbers than other bee species and have been documented to pursue perceived intruders a quarter of a mile. Some 1,100 human deaths have been attributed to their stings, and they have migrated steadily northwards, reaching Texas and California by the early 1990s.[15]

Animal Organ Farm

Would humans in the rich world with access to advanced healthcare start to diverge physiologically from those in low-income countries? Alternatively, could a black market of underground gene-editing clinics for athletes, entertainers and the affluent seeking an edge for their offspring start to emerge in less regulated parts of the world?

As pointed out earlier, there was a global outcry when a Chinese scientist, He Jiankui, claimed to have created the world's first genetically edited babies in 2018. The Chinese Government prosecuted and imprisoned him.[16] Similarly, the World Health Organization and Moscow blocked a plan by a Russian biologist Denis Rebrikov in 2019 to use CRISPR to help deaf parents with an inheritable condition to have children with normal hearing ability.[17]

Synthetic biology and gene editing also have implications for organ, tissue and cell transplants. Worldwide, patients die everyday while on wait lists for organ transplants. We have the technology to change that, but the implications of just how are unsettling to contemplate.

In his 2005 dystopian science fiction novel *Never Let Me Go*, British novelist Kazuo Ishiguro imagined an England in which children are cloned and bred to harvest their organs. It's a safe assumption that most would find that morally abhorrent.[18]

Yet what if you swapped out children for pigs? The demand for organs worldwide has accelerated research in xenotransplantation, or the transfer of live cells, tissues and organs to patients from non-human sources. Pigs have organs similar in size and physiology to that of humans.

Cross-species transplantation is enormously difficult, due to the risk of graft rejection by our immune system. There's also the threat of latent retroviruses and infectious disease transfer after a procedure.

However, a spin-off company from Harvard University biologist George Church's lab called eGenesis has made progress engineering the genomes of pigs to make their cells more compatible with the human body using CRISPR gene-editing technology.

Other researchers are exploring ways to genetically modify pigs to mitigate those threats.[19] If they're successful, humanity could have a robust supply of organs, heart valves and other body parts to service human health. In early 2022, a U.S. man became the first patient ever to get a heart transplant from a genetically altered pig, though he died soon after. Animal rights activists want to phase out livestock slaughter for meat consumption. What if the swine killing saves human lives? Is that a different matter?

It's hard to know if the taboos of the 2020s about designer babies and genetically manipulating ecosystems will still be around in the 2050s and beyond. The temptation to use human gene editing and synthetic biology for military advantage may be irresistible for leading world powers. U.S. intelligence officials claim the Chinese are conducting human tests on members of the People's Liberation Army to bio-enhance their soldiers.[20]

A U.S. Defense of Department study, meanwhile, predicted that by 2050 the American military would be able to improve the performance of its soldiers with neural implants for two-way communication with weapons systems and reconnaissance drones.[21]

Scientists with the U.S. Army Research Laboratory are studying how skin bacteria might be genetically modified to improve military body armour and how to modify the microbiomes of soldiers so they perform better in stressful conditions.[22]

There's also an economic race to dominate life science technologies in the century ahead. Right now, the U.S., United Kingdom, the EU and Japan are leaders. China is investing huge sums into research and development, and the quality of its scientific research has rapidly improved over the last two decades.

The Shenzhen-based BGI Group, which makes genetic sequencing equipment, diagnostic tests and handles research for drug companies is emerging as a globally competitive Chinese life sciences player. Such competitive pressure suggests that scientific powers have every incentive to keep pushing ahead, rather than tapping on the brakes.

The World Health Organization, the U.S. National Academies of Sciences, Engineering and Medicine and the U.K. Royal Society have all issued reports suggesting guidelines for this uncertain era we are moving into. Establishing red lines such as heritable gene editing that can be passed onto offspring, the creation of scientific registries to track research work, setting up oversight committees to review sensitive research and human clinical trials, and whistle-blower protections are among the proposals being floated.

It's a historic, if somewhat fraught, moment in the history of our species. Humanity is no longer just a passive passenger in the four billion-year flow of evolutionary life. We're now creators of life, too.

Biology at the molecular level is a complex and self-replicating dynamic. Mistakes may be passed down to future generations for good or for ill. A premeditated bioterrorism attack by malicious actors can do incalculable damage and might be difficult to reverse.

So that leaves us with the paradox of synthetic biology. It's both a wondrous platform for discovery and progress, but it also might end up doing a lot of damage to human civilization. Managing biorisk in this new age will take sustained effort, smart policies, and vigilance.

Chapter 13

Biorisk Without Borders

Andrew Weber has spent much of his career thinking about biorisk in a rapidly evolving landscape. As a young American foreign service officer, he helped uncover a secret biological weapons program in Kazakhstan during the 1990s. Now, he ponders a new era in which biological computational tools and desktop DNA synthesizers pose far different biosecurity challenges.

"We are no longer talking about Russian biolabs with 20,000-liter fermenters," Weber told me in a video interview from his home. "It really is possible to develop enough agents to start a pandemic, or kill millions of people, in a small house. As it all goes digital, people could recreate the smallpox virus. The recipes are all publicly available," on the Internet.

Roughly three decades ago, in the aftermath of the Soviet Union's collapse, control over one of the world's largest arsenal of nuclear weapons fragmented among newly independent states, much to the unease of weapons experts in the West.

In 1995, Weber and his team unmasked an offensive Russian biological weapons program and a massive anthrax production facility in Stepnogorsk, a city in Kazakhstan that didn't feature prominently on many maps at the time.[1] Weber, a senior fellow at the Council on Strategic Risks, a national security think tank in Washington, D.C., believes the lack of awareness in the West about Russia's biological weapons ambitions was one of the most significant U.S. intelligence failures during the Cold War.

Weber recalled his astonishment when he walked through a 200-yards-long Soviet biological weapons production center, equipped with four-storey-high bioreactors that had the capacity to produce 300 metric tons of weaponized anthrax in about eight months. The fermentation area was a maximum-security biolab with workers wearing space suits hooked up to oxygen lines.[2]

The Soviets had also refashioned an island called Vozrozhdeniya in the Aral Sea into an open-air, biological testing facility for super-pathogens. Researchers exposed monkeys and other animals to toxins, and then took them to high-containment labs to monitor disease progression.

"That was surreal," Weber recalled. "They would have over 800 soldiers and scientists there for months on end. It was incredible that they would test these most horrific agents in open air."

In the Obama Administration, Weber later served as Assistant Secretary of Defense for Nuclear, Chemical and Biological Defense Programs and played a lead role in shaping the U.S. response to the Ebola outbreak in Africa.

During Weber's time at the Pentagon, advances in genetic sequencing and DNA synthesizing had raised disquieting questions about emerging biosecurity threats. "What would these new capabilities to write and edit DNA and program cells give our adversaries?" Weber and his team wondered. In addition, these dual-use technologies were also becoming widely available to industry actors and academic researchers, another layer of potential proliferation risk.

Synthetic biology, an engineering-focused scientific discipline emerging from the convergence of biotechnology, genetics and advances in microbial science, has deepened the world's understanding of how living cells function. New tools have emerged for redesigning living systems at the genetic level, or even creating entirely new microorganisms. Along with the potential benefits for new drug discovery, consumer products and the emergence of a more sustainable bioeconomy came new types of biosecurity risks.

International biodefense arrangements have focused primarily on deterring state actors. Chief among them is the Biological Weapons Convention, an international treaty that took effect in 1975 and prohibits the development, production, acquisition, transfer, stockpiling and use of biological and toxin weapons.[3]

Signed by more than 180 states, it was the first multilateral disarmament treaty banning an entire category of weapons of mass destruction (WMD). In addition, United Nations Security Council Resolution 1540 prohibits governments from assisting non-state groups from securing or developing biological, chemical, radiological and nuclear weapons.[4]

In 2010, the U.S. Department of Health and Human Services issued industry guidelines for the then-fledgling synthetic biology field to limit the dual-use technology risk and suggested ways to restrict access to researchers and clients with the appropriate backgrounds and credentials. Companies that specialize in DNA synthesis voluntarily screen orders for sequences from lists of regulated pathogens and potentially toxic biological material maintained by various governments.

The International Gene Synthesis Consortium (IGSC), an industry-led organization focused on sharing best practices in biosecurity screening, encourages its members to flag orders for synthetic DNA from the causative agents of smallpox, anthrax and rinderpest, among other diseases, for scrutiny and review.

Suspicious requests to synthesize DNA can be compared against international sequence and organism databases such as the MicrobeNet, GenBank and the Multidrug-Resistant Organism Repository. The trade industry group is also working on a biosecurity screening accreditation program.

A 2018 study by the National Academies of Sciences, Engineering and Medicine (NASEM) by leading experts in biotech and synthetic biology and funded by the U.S. Department of Defense found vulnerabilities in the current biosecurity regulatory framework. It concluded that "approaches modeled after those taken to counter Cold War threats are not sufficient to address biological and biologically enabled chemical weapons in the age of synthetic biology."[5]

In America, the government's Federal Select Agent Program oversees the possession, use and movement of well-known pathogenic microbial strains and toxins.[6] That may not be comprehensive enough protection in an age in which a potential terrorist could design an entirely new infectious virus or tweak an existing one extensively enough to evade detection. "Such approaches will not be effective in mitigating all types of synthetic biology-enabled attacks," the NASEM study pointed out.

Weber thinks the whole approach to biosecurity in the U.S., not to mention worldwide, needs a serious rethink. "It's a very fragmented and weak system, and we need to find ways to strengthen it to make the development and use of biological weapons taboo," he explained.

While it remains technically challenging to weaponize biological agents, someone with the right level of expertise and access to equipment could do serious damage. Weber thinks the world missed the larger significance of the 2001 Amerithrax attack, in which an American government biodefense scientist, Bruce Ivins, allegedly sent letters full of anthrax to media outlets and politicians.[7]

As pointed out earlier, Ivins, who worked at the U.S. Army Medical Research Institute of Infectious Diseases, committed suicide before being charged by the Federal Bureau of Investigation in the bioterrorism attack, which killed 5 and sickened 17, in the worst domestic biological attack in U.S. history.

"He had the quality and quantity of anthrax that could have killed tens of thousands of people if that had been his intent," according to Weber. "Once you are exposed to inhalational anthrax, if you don't know that until you're displaying symptoms four to seven days later, then it's very, very hard to treat with antibiotics. The lesson for me is that one person can mount an enormous attack."

Grim Precedents

Biowarfare has been a chilling reality since the fourteenth century, when the Mongol Army catapulted plague-ridden bodies over city walls during the siege of the Crimean city of Caffa.[8] World War I is well-known for its use of mustard and chlorine gases. Yet biological weapons made an appearance, too, when the Germans transported horses infected with anthrax behind enemy lines in a bio-sabotage attack.[9]

In World War II, the U.S., United Kingdom, Canada, Germany, the Soviet Union and Japan all had active biological weapons programs. The Japanese Imperial Army's Unit 731 used prisoners of war in China to test pathogens and dropped bombs containing plague-infected fleas on Chinese cities.[10]

In 1962, during the Kennedy Administration, the U.S. started a bioweapons research program called Project 112 that over the years would expose American servicemen and the public to chemical and biological agents.

The Defense Department in 2003 identified more than 6,200 individuals who were potentially exposed before the program shut down in 1973, according to a U.S. Congressional report.[11]

Flash forward to the 2020s, and the biorisk environment for governments is far different. We live in an age of outsourced DNA synthesis work, automated "cloud" labs and a global movement of amateur biohackers, some of whom enthusiastically advocate human enhancement through redesigned microbes, biological cells and implant technology.

Under existing rules, two Canada-based virologists in 2018 had little difficulty getting a mail-order synthetic biology company to synthesize the DNA of an extinct horsepox virus.[12] That kind of work is completely legal, despite the fact it has applications for developing a human killer like smallpox.

DNA sequencers and synthesizers basically allow us to read and write the source code of life using the four nucleic acid bases: As, Cs, Gs and Ts corresponding to adenine, cytosine, guanine and thymine.

Over the decades, the technology has become orders of magnitude faster and cheaper. Companies such as Illumina have built up a great business selling DNA sequencing machine and services, as have firms such as Twist Bioscience that produce synthetic DNA on demand after screening clients.

Now, companies such as French biotech startup DNA Script have developed a desktop DNA sequence and assembly device, or "printer," that could be used anytime, anywhere. That's going to distribute an awful amount of technological power into the hands of researchers in labs worldwide. The vast majority of scientists act responsibly of course. However, DNA synthesis work is done all over the world. And there's still not yet a global system in place to prevent bad actors from doing suspicious research.

It has fallen to individual companies to fill the gap. Twist Bioscience has built robust systems and processes for biosecurity screening of orders to produce synthetic DNA that it receives from clients worldwide.

Based in South San Francisco, Twist has developed a DNA manufacturing system that's helped change the economics of synthetic biology. In past decades, synthesizing DNA used to involve an intricate, time-consuming process of mixing chemicals and reagents on plates the size of an index card.

Founded in 2013 by Emily Leproust, Bill Banyai, and Bill Peck, Twist found a way to miniaturize the process using a silicon-based plate with tiny compartments and an automated system that can dispense miniscule chemical droplets of 10 picoliters. A picoliter is a millionth of a microliter, which is one millionth of a liter. We're talking small here.

This means Twist can make millions of pieces of DNA sequences at far less cost. Its research clients can also do more experiments to unravel the biological mysteries they're studying. Big breakthroughs in drug discovery or consumer products are more likely to happen if scientists can explore more genetic sequences and variants of their research idea.

James Diggans, who leads biosecurity at Twist and earned a PhD in computational biology and bioinformatics from George Mason University, gets paid to think about what could go wrong. The risks are real. However, Diggans thinks science writers and the public generally don't always grasp the enormous technical challenges involved in turning a microbe into a weapon of mass destruction.

"It's easy to tell scary stories about biology," Diggans explained, "and it's really hard to build an organism to do anything you want it to do." A smudge of anthrax DNA at the bottom of a test tube is pretty far upstream from a direct weapons threat. "You still have to scale that up. You have to figure out a way to disperse it," according to Diggans.

The fast-evolving world of synthetic biology and the widening access to its tools doesn't automatically increase the risk of weaponization, in his view. "But it does increase the complexity of the landscape for biological functions that a company like Twist would want to be on the lookout for," according to Diggans.

His team screens clients and will have conversations if the research appears sensitive or potentially dangerous. In some cases, Twist customers simply haven't fully understood the risks of the microbes and biological agents they were working with and dropped their projects after being alerted by Diggan's biosecurity team.

Diggans thinks the U.S. and international biosecurity framework needs to be better defined, more global and better financed with government help. He's also adamant that synthetic biology companies need to consider biosecurity as part of the cost of doing business, rather than a nice-to-have capability.

In addition, science educators should include biological risk assessment in their curriculums. "That's not generally how molecular biologists are trained. Evaluating the security implications of research is a skill set that in the future is going to be more and more important. It should certainly be part of science education more broadly," he said.

Diggans believes it's time to shift the regulatory framework away from a focus on the species of microorganisms whose genes are sequenced to the biological function of the research involved. In other words, a bad actor could try to synthesize DNA that might not map exactly to a pathogen on a government infectious disease control list, but still be potentially dangerous to society.

Another issue is the screening parameter for the length of gene sequences that need to be scrutinized under the 2010 U.S. Government guidelines. DNA synthesis often involves creating short fragments of DNA called oligonucleotides, or oligos, and then using enzymes to sew them together. Right now, any sequence under 200-base pairs of DNA generally isn't required to be evaluated in a biosecurity screen.

A lot has changed in the ensuing decade, however. Twist and other DNA synthesizers can now work with vast pools of oligos. In theory, a lot of tiny gene sequences that look benign in isolation, could be more concerning stitched together into longer segments.

"These are very large pools that include a lot of biological complexity," according to Diggans. "You want to have the ability to make sure that those pools can't be used to assemble something that we wouldn't otherwise be willing to sell to a customer. It shouldn't represent a route for evasion of biosecurity controls."

Further out, biosecurity experts wonder what the world would look like if DNA synthesizers became mobile and sophisticated enough to be scattered in labs all over the world. That kind of widely distributed synthetic biology manufacturing will be much harder to manage than centralized corporate providers like Twist.

"If you are going to have a device that's capable of synthesis of DNA and it's on a benchtop in someone's lab, you need to make sure that device has the same kinds of controls associated with it that a centralized provider would," notes Diggans. "If I ask that machine to spit out smallpox, it had better say no."

Diggans and others believe that software layers and AI algorithms can improve biosecurity. Powerful computer programs such as

AlphaFold that can predict the structure of proteins in 3D that have been developed by Google's AI subsidiary DeepMind could be potentially transformative in Diggans' world.[13] A protein's shape impacts its function. Biosecurity experts could train AI programs to predict possible protein structures for ordered DNA sequences and then be on the outlook for risky research.

Biosecurity is expensive, and for start-ups that can be a challenge. To help smaller companies cope, a security expert group called the Nuclear Threat Initiative is working on an open-source screening mechanism.

Diggans also hopes the U.S. Government will research and develop a database of biological threats at the sequence level that can be shared privately with the industry, similar to the information flow about cyberthreats between the information tech industry and Washington. The better the biorisk intelligence from around the world, the more often biosecurity screening software can be updated.

"Export control says that any sequence that can endow or enhance pathogenicity from a listed organism requires an export licence. Which sequences enhance pathogenicity? There's no list for that," according to Diggans. "So it makes every incoming sequence order a mini-science project."

Ginkgo Bioworks co-founder and synthetic biology godfather Tom Knight also thinks we need deep databases of potentially dangerous genetic sequences that researchers and DNA synthesizing companies can refer to as they consider projects and genetic modification work. In his view countries should be doing extensive testing in all manner of heavily trafficked public spaces to detect potential microbial threats.

"The U.S. Government should be all over this, something akin to the National Reconnaissance Office that designs, builds, launches and maintains America's intelligence satellites. It should be out there sampling everything," according to Knight.

Right now, there's a list of organisms and toxins that scientists and experts can refer to in the Australia Group, a multilateral export control forum that includes the U.S., Canada, and the U.K., as well as other nations in Europe and Asia. Knight doesn't see China or Russia ever joining that kind of effort, but hopes they will build their own biobanks of data as a biosecurity precaution in those parts of the world as well.

Smallpox.com

A tougher issue for free societies is what, if anything, to do about the tremendous amount of open-source information that can be quickly accessed about the genetic sequences of microbial pathogens, DNA assembly know-how and biological components that perform specific functions.

Complete or partial genetic sequences of the deadliest human pathogens that cause anthrax, the plague, and smallpox have been posted on the Internet. It's not hard to find them with a search engine.

With certain pathogens like *Bacillus anthracis*, the naturally occurring bacterium that causes anthrax, an aspiring bioterrorist with the right expertise and equipment wouldn't even have to bother breaking into a lab for a live sample. The spore-forming microbe can be found in the soils of Texas cattle country, where livestock and deer are regularly infected after grazing in contaminated areas.[14]

It would require considerable sophisticated know-how, but the technology now exists for a bioterrorist with the right training and access to lab equipment to genetically modify an existing Avian virus to be super-transmissible or bacteria to resist antibiotics.

Consider, too, that in the age of synthetic biology, the distinction between chemical weapons and biological weapons all but disappears. Microbes such as bacteria and fungi produce toxins in nature for defensive purposes that are quite lethal to humans.

Botulinum neurotoxins that block nerve cells and can cause respiratory and muscular paralysis are made from the bacterium *Clostridium botulinum* and pose a major bioweapon risk, given its potency and lethality, particularly if aerosolized. In theory a single gram of the toxin, evenly dispersed and inhaled, could result in more than one million fatalities.

During the 1980s, Iraq maintained a robust biowarfare research program, according to revelations by United Nation inspectors after the first Gulf War in the early 1990s. At a production facility in Al Hakum, about 60 kilometers southwest of Baghdad, scientists weaponized botulinum and aflatoxin, a toxic chemical produced by fungi.[15]

From the perspective of an aspiring terrorist, biological weapons offer advantages over conventional chemical and nuclear weapons. The ingredients of bioweapons are microorganisms that often replicate inside the host, so they pack more lethality per unit of weight and can be stockpiled in smaller levels than chemical weapon agents.

Whereas production of chemical weapons requires precursor materials and process equipment that's easier to trace, bioweapons can be made from dual-use machines and in legitimate research settings.

True, there are also some drawbacks. Amateur biowarriors can get themselves killed handling pathogenic microbes if they don't know what they're doing. Delivering these biological agents to large population centers is also exceedingly difficult.

Drones and agricultural sprayers on airplanes might work, but shifting winds make precise targeting near impossible. Long-range delivery systems like cluster bombs require the kind of military resources that only governments can muster.

That said, according to the 2018 NASEM report, advances in biotech and synthetic biology have made it far easier to produce militarily meaningful batches of pathogens and toxins and come up with diabolical ways to trip up government responses after an attack.

For instance, a viral or bacterial pathogen could be programmed to first infect populations with a mild form of a disease that could spread broadly, before evolving into a much more lethal strain that would then overwhelm healthcare systems. Right now, the report sees that as a low probability risk, but such scenarios are on the minds of biosecurity experts.

A sophisticated adversary could also introduce bacteria into a country's food chain that might quietly unbalance the gut microbiomes of large numbers of people, causing widespread illness and undermining immune systems. Given its relative ease of use and low cost, gene-editing systems like CRISPR were flagged for the first time in America's annual threat assessment report by the U.S. Director of National Intelligence in 2016.[16] "Its deliberate or unintentional misuse might lead to far-reaching economic and national security implications," the report noted.

Bioterrorism doesn't have to be directed at humans to be strategically effective. Sickening livestock and poisoning agriculture to undermine food security would cause widespread panic and cripple economies. In China, criminal gangs have used drones to drop products contaminated with swine flu into pig farms as part of an extortion racket to force farmers to sell their animals at a steep discount.[17]

Disinformation networks designed to undermine national responses to biological events is yet another form of biowarfare. Intelligence agencies say Russian disinformation networks had fabricated vaccine scare stories, amplified by social media networks like Twitter and Facebook, years before the COVID-19 health emergency to stoke anti-scientific sentiment and deepen political polarization in the West.[18] The effort reaped dividends when the pandemic swept over the planet.

The U.S. military has long factored in the risk of biological weapons into its war games. In late 2020, as the SARS-CoV-2 virus tightened its grip, the U.S. Air Force simulated a conflict with China set in the 2030s. The Chinese executed a successful biological-weapon attack that ran riot through U.S. bases and warships in the Indo-Pacific region. Next came a Chinese missile strike, followed by an air and amphibious assault of Taiwan. In this simulation, American forces were soundly routed.[19]

Biohackers in Arms

Aside from state actors, biosecurity policymakers need to contend with a worldwide biohacking movement, also called DIY biology, that is characterized by different ideological strands and participants ranging from PhD-trained biologists to amateurs.

For some, the biohacking creed is about enhancing our brains and bodies with gene editing and synthetic biology. For others, it's about taking control of one's biological destiny by reprogramming the inherited gene set from their biological parents.

Still others are fascinated by the technological possibilities of wearable devices to track health metrics, neurofeedback to understand brain rhythms and human-machine body implants to monitor glucose or enhance human intelligence.

Some biohackers are avid transhumanists, a cultural and political movement that sees futuristic technologies like genetic modification, cyborg body enhancement and even mind and memory uploading profoundly changing the human species.

One of its leading lights, Zoltan Istvan, ran unsuccessfully for U.S. president in 2016 as head of the Transhumanist Party. Driving around the country in a campaign bus shaped like a coffin, he advocated mandatory college and universal basic income to prepare Americans for a society run by robots.

The growing sophistication of amateur garage biologists raises the risk of the accidental release of dangerous microbes into the environment. In the U.S., the FBI and the American Association for the Advancement of Science have done outreach programs to DIY biohackers to raise awareness of biorisks from synthetic biology and gene-editing technologies.

During the COVID-19 pandemic, a Chinese lab facility at the Wuhan Institute of Virology came under scrutiny from the Western media, intelligence agencies and social media conspiracy theorists. Might have SARS-CoV-2 accidentally leaked from the Wuhan lab?

The maximum biosafety level 4 lab (BSL-4) in Wuhan did study coronaviruses taken from bat samples in central China. One speculative theory is that the virus was among the samples collected by lab researchers in the field. Then, an infected lab employee unknowingly passed the contagion to the broader community, setting in motion a pandemic that would kill millions, according to this scenario.

There's no conclusive evidence COVID-19 emerged from a lab. What has triggered suspicion is China's unwillingness to completely open up its Wuhan biolab, staff and records to international scrutiny. Also arching eyebrows: Starting around 2016, researchers at the Wuhan biolab had conducted experiments on a bat coronavirus called RaTG13 that's 96.2 percent genetically similar to the COVID-19 virus SARS-CoV-2.[20]

The U.S. State Department claims the lab altered, then removed online records of its research of RaTG13 and that several researchers at the Wuhan institute became sick in the autumn of 2019 with symptoms similar to COVID-19 and common seasonal illnesses.

Ironically, the Wuhan lab indirectly received funding from the U.S. government. Among the Chinese virology institute's research partners was EcoHealth Alliance, a New York-based global health group that received a research grant from the National Institutes of Health for infectious disease studies. The Trump Administration cut off funding in 2020, though it was later reinstated with strict and numerous conditions.

One of the lab's virologists is Shi Zhengli, known by public health professionals in China as the "bat lady" for her virus-hunting excursions into bat caves. After the viral outbreak in Wuhan, Shi and her colleagues ran tests of the genetic sequences of samples from her lab and infected patients in the general population. None of the sequences matched, Shi told *Scientific American* in an interview.[21] Shi has also claimed there was "zero infection" among the WIV's staff.

It's unlikely the mystery over the origins of COVID-19 will ever be resolved without greater transparency from Beijing. Yet forget Wuhan for a moment. There are plenty of reasons to be concerned about the growing number of maximum security, or BSL-4 labs, popping up across the world, according to biosecurity experts.

There are about 59 containment labs operating, or planned, that handle some of the deadliest pathogens found on Earth. At some of these labs, scientists perform "gain of function" experiments to observe changes in a microbe's genome for research purposes, some of which may coincidentally result in the increase of a pathogen's transmissibility or lethality.

Advocates say such work will help drug developers design better treatments. Skeptics worry about the unintentional release of a super-strain of influenza or bacteria that kills millions and destabilizes the world.

Gain of function research drew criticism in the early 2010s, after researchers in Wisconsin and the Netherlands received funding to genetically modify a lethal Avian flu virus to enhance its transmissibility.[22]

Soon after, there was a series of lab accidents at the U.S. Centers for Disease Control and Prevention involving mishandled pathogens.[23] In one instance, there was an accidental shipment of live anthrax; another mishap involved the discovery of forgotten smallpox samples that were still metabolically active.

In late 2014, the NIH announced a pause on U.S.-funded gain-of-function research that involved influenza, severe acute respiratory syndrome and the Middle East respiratory syndrome.[24] The institute later allowed most, but not all, of these studies to go forward after review.

The maximum security BSL-4 labs span 23 countries with contrasting levels of regulatory oversight. Two biosecurity experts, King's College London senior lecturer Filippa Lentzos and Gregory Koblentz, an assistant professor at George Mason University, discovered that not all of the labs scored high marks in a Global Health Security Index that measures regulatory oversight and biosafety training.[25]

Roughly one quarter of the countries with BSL-4 labs received high marks for biosecurity preparedness. About one-third of the countries, China included, had medium levels. A facility in South Africa received low marks.

Accidents of varying degrees of severity are regular occurrences at biolabs worldwide. In 1978, in what's believed to be the last known case of smallpox, a medical photographer named Janet Parker contracted and died from the disease.[26] She apparently was exposed to the variola virus at the smallpox laboratory at the Birmingham Medical School where she worked.

In 2004, two Chinese researchers unknowingly infected by SARS caused a mini-outbreak in Beijing.[27] In 2019, the U.S. Army Medical Research Institute for Infectious Diseases in Fort Detrick, Maryland, shut down over concerns about possibly contaminated wastewater from its high-security labs.[28]

Nor are all the countries with maximum security biolabs members of the International Experts Group of Biosafety and Biosecurity Regulators, in which national regulatory authorities share safe and effective lab management practices.

"The vast majority of countries with maximum containment labs do not regulate dual-use research, which refers to experiments that are conducted for peaceful purposes but can be adapted to cause harm; or gain-of-function research, which is focused on increasing the ability of a pathogen to cause disease," Lentzos and Koblentz concluded.

SynBio Sentinels

One way to improve our situational awareness of new pathogens, whether accidentally leaked or maliciously designed, is to supplement the work of human biosecurity analysts with artificial intelligence.

The U.S. Government and synthetic biology companies are developing AI software tools that can recognize statistical anomalies in DNA segments, equipment usage or research requests related to a pathogen that might be related to an emerging biological threat.

Algorithms that sift through biological databases might be able to distinguish between genetic sequences that seem to have been engineered versus occurring naturally.

In the U.S., companies are working with the Intelligence Advanced Research Projects Activity, an agency with the Director of National Intelligence office, on an AI-powered software program to monitor DNA synthesis. Software programs are being designed to identify whether a particular DNA sequence was engineered and by whom, as well as its potential risk.

AI applications have the potential to improve biosecurity only if the bots have access to comprehensive and standardized data sets that cover the entire process of synthetic biology from design to finished product.

More work needs to be done by governments and the life sciences to track lab work, personnel and supply chains. A digital signature system of cell designs and production might deter bad actors, as would more robust methods that track design software files and the work of automated Synbio foundries.

Furthermore, there is an opportunity for screening procedures to move from a focus on organisms to biological functions. Biosecurity software programs and databases need to be constantly updated with the latest developments from synthetic biology labs.

Aside from AI tools and counter-terrorism programs by governments, the best defense against bioterrorism involves the same methods explored earlier in this book about pandemic preparedness. We need to invest in surveillance systems that can identify infectious pathogens early.

Molecular maps of the microbial landscape, flexible and fast vaccine platforms, AI-powered algorithms scanning for early signs of an outbreak, all could play a role in not only lowering the chance of another pandemic but also possibly deterring bioterrorists.

Reducing the catastrophic risk of a biological attack is a game of probabilities. We will never develop a full-proof system. That said, we can definitely increase the odds of avoiding a bioterrorist attack. Investment in a global digital network to track sensitive synthetic biology work, AI tools and information sharing among countries will be far less expensive than the casualties and hit to the global economy that a successful attack would cost.

Chapter 14

Microbial Eve: A Creation Story

How did it all begin? Mapping out how self-replicating, biological life sprung into existence on Earth is one of science's greatest endeavors, requiring a convergence of biology, physics, geology, chemistry, and cosmology. It's a mystery that, when unraveled, would mark another transition point in humanity's understanding of its place in a vast, complex and interconnected universe.

The search for the origin of life is simultaneously an exploration into our primordial past and deep future. Was the chain of events that sparked life on Earth a random fluke? Or is life probable, even inevitable, given the right geochemistry and physical processes on other planets?

As the dominant life form during the planet's early history, microbes have revealed a great deal about the early progression of proto-cells to modern living cells billions of years ago and pose theoretical questions, whose answers may have big implications for the future of our species.

The current scientific consensus is that life — as in organized cellular matter that consumes energy, responds to its environment, grows, reproduces and evolves over time — started somewhere between 3.5 billion and 4 billion years ago.

Paleobiologists have been combing through ancient rock formations around the world, including a deposit called the Apex Chert in Western Australia. They've excavated, pulverized and lasered-blasted rocks looking for microbial microfossils and evidence about when life got going on the early Earth and under what kind of chaotic conditions.

In 2017, a team of scientists from the University of Southern California and University of Wisconsin–Madison reported that their carbon-dating analysis of the oldest known fossilized microorganisms revealed they were 3.465 billion years old. Two of the species analyzed in the study performed a primitive form of photosynthesis, while another produced methane gas. Two others consumed methane and appeared to have used it to build their cell walls.[1]

Other research teams claim to have found evidence of microbial life as far back as 3.7 billion and 4.1 billion years ago, though those findings, and the chemical analysis underpinning them, are hotly contested by experts.

Some geological studies point to Earth's crust having cooled and solidified as far back as 4.4 billion years ago, and water may have accumulated 100 million years after that.[2] Aside from the research about timing, just exactly *how* life started is also the subject of an animated scientific debate, with several competing theories in contention.

The discovery of exoplanets orbiting around stars outside of our solar system starting in the early 1990s renewed interest in origin of life studies. Might there be a discernible path from non-life to life, from physics and chemistry to biology, that could be replicated under the right conditions anywhere in the universe?

The more we understand how life can emerge from nonliving matter, a process called abiogenesis, the more informed we will be in our search for extraterrestrial life. Such knowledge may be crucial if humanity is to have a realistic shot of creating a truly multi-planetary civilization.

The shift in our perception about the fabric of an expanding universe has changed radically since the start of the twentieth century. The current consensus among most astrophysicists is that our universe started about 13.7 billion years ago after energy that was packed into an extremely dense point exploded with unimaginable force.

The matter created by this Big Bang radiated outward, eventually creating galaxies and planets. Earth arrived on the cosmic scene about 4.5 billion years ago as a swirling cloud of gas and dust left over from the formation of the sun condensed into a super-heated planet.

The basic chemical building blocks (hydrogen, helium, carbon, nitrogen, sulfur, and phosphorus) needed for the possible emergence of biological life are believed to have existed in one form or another in the interstellar space of the early solar system.

Yet early Earth, far from a nurturing cradle of life, was a turbulent cauldron of molten rock — hence, the name Hadean Eon after the Greek god of the underworld called Hades. Massive impact collisions with extraterrestrial bodies unleashed enormous amounts of heat that destabilized the core of the planet and prevented much rock from coalescing at the fiery surface.

The Moon is believed to have been formed during this period, possibly the result of a collision between Earth and a rocky body the size of a small planet that, in turn, ejected material into our gravitational orbit. The early atmosphere consisted of hydrogen and helium initially, then with the rise of a volcano system came carbon dioxide, hydrogen sulfide, ammonia, methane and neon. Our oxygenated atmosphere wouldn't arrive until much later, about 2.4 billion years ago, when cyanobacteria evolved to perform photosynthesis, converting carbon dioxide and solar energy into oxygen.

Eventually the bombardment of space rocks from the inner solar system began to subside and Earth's surface cooled. While there's no clear consensus, some researchers claim our primitive oceans started to form as water vapor escaped from molten rocks in a cooling world.

When Earth's temperature fell below the boiling point of water, the vapor condensed into rain that collected in the hollowed crevices of the surface over millions of years. Other scientists speculate that water-carrying comets and asteroids that slammed into Earth in the late Hadean Eon may have contributed to water resources.

Warm Little Pond

Darwin, whose theory of evolution outlined in *On the Origin of the Species* caused an uproar in Victorian England, suggested in a letter to a colleague that life might have started in a "warm little pond" with chemicals mixing with sunlight, heat and electricity to create a protein compound.[3]

In the 1920s, the notion was developed further by Soviet biochemist Alexander Oparin and British biologist John Burdon Sanderson Haldane. Both independently speculated that the various chemicals swirling in

Earth's early oceans might have reacted with each other, forming complex compounds that over time evolved into primitive cells.

The concept of a primordial soup of non-living chemicals evolving into early life forms became known as the Oparin-Haldane hypothesis. Then, in 1952, chemistry Nobel laureate Harold Urey, a key scientist involved in America's crash, wartime Manhattan Project to build an atomic bomb, conducted an audacious experiment with a young doctoral student, Stanley Miller.[4]

In an attempt to recreate the conditions of early Earth, the scientists mixed together ammonia, hydrogen and methane gas and boiling water. They shocked the mixture with electricity to imitate a lightning strike. The simulation triggered the formation of amino acids, the building blocks of protein and by extension biological life.

A year later, Francis Crick and James Watson with the University of Cambridge worked out the inner workings of the double-stranded DNA as a storehouse of genetic information essential for biological life.[5]

They explained how our genes replicated and passed hereditary information down from parent to offspring during the history of our species and speculated that all genes were ultimately traced back to an ancestral microbe. Before that first microbe arrived, though, some sort of primitive cell had to have emerged. Scientists have spent decades trying to crack the code but have no definitive answers yet.

Most cells that make up all living things rely on basic information-carrying molecules. One is DNA (deoxyribonucleic acid) that encodes the genetic instructions or blueprint for biological life. Its simpler, one strand relative, RNA (ribonucleic acid), converts that code into proteins made up of amino acids that get stuff done. Proteins are essential to biological processes, constructing and repairing an organism's parts and overall structure, while also serving as an energy source. Finally, encasing and holding everything together is a membrane composed primarily of fatty acids and proteins.

There's no broad consensus about how proto-cells formed. But somehow they did, and the resulting self-replicating molecule eventually mutated and evolved into the first primitive living cell, probably a bacteria or something very close to one. Geneticists and evolutionary biologists call it the Last Universal Common Ancestor, or LUCA, from which all life descended.[6]

Historically, the tried and true metaphor that captures the evolutionary dynamic through time has been the tree of life, with a trunk extending upwards into a complex network of branches. Darwin introduced his notion of a tree of life in his masterwork *On the Origin of Species* in 1859.[7]

Seven years later, a German zoologist named Ernst Haeckel really embraced the idea, producing a comprehensive tree that extended over 1,000 pages, based on the knowledge of the day. In Haeckel's view, there were three major branches: Plantae, Protista and Animalia.

While many of his classifications have long since been refined or discarded, Haeckel is credited with creating core concepts of evolutionary biology, starting with the idea that organisms could be sorted systematically by their characteristics. (Far less admirable were his contributions to social Darwinism and scientific racism.) Thus was born the field of phylogeny, or the study of the evolutionary history of species via lines of descent and relationships between groups of organisms.

Tangled Tree of Life

A century later, the quest to understand the evolutionary history of life would get one of its most innovative thinkers in Carl Woese, who had received a PhD in biophysics at Yale. After a brief stint as a researcher at General Electric, Woese headed to the University of Illinois Urbana-Champaign in 1964.

Woese was one of the more eccentric, and indisputably brilliant, figures in twentieth-century biology. A sometimes enigmatic man, he's credited with rewriting the history of biology with his discovery of a new domain of life called Archaea. These microbes tended to live in extreme environments and were molecularly different enough from bacteria to warrant a new classification of biological life.

He preferred flannel shirts and jazz bars (he kept a giant poster of Miles Davis in his office) to tweed jackets and faculty lounge cocktail hours. Nor was he particularly skilled at, or interested in, navigating the political waters of the academic world, where perceived slights occasionally escalate into career-derailing feuds.

Woese didn't spend a lot of time ingratiating himself with university administrators who controlled department budgets. He was famous for abruptly ending lunches with his superiors when he felt there was nothing more to say. His sense of humor was idiosyncratic, rambunctious and sometimes biting.

"Carl had a sub-clinical Asperger's type of personality, and this meant that he was unable to put himself in the mind of the other person," wrote University of Illinois physics professor Nigel Goldenfeld, a close friend and collaborator with Woese for many years.[8] "His humor was, thus, sometimes unfettered by good taste; but it always betrayed the fact that he looked at life in a very different way than did other people."

The unique sensibility animated his research interests. Biologists had long sorted microbes by their shapes, or morphologies, and behaviors such as how they metabolized food into energy. With his background in physics, Woese searched for a quantitative way to journey back through time to LUCA, the ancestral bacteria from which all life sprang.

In a letter to Francis Crick, the co-discoverer of DNA, Woese wrote in 1969 that, "If we are ever to unravel the course of events leading to the evolution of the prokaryotic (i.e., simplest) cells, I feel it will be necessary to extend our knowledge of evolution backward in time by a billion years or so; i.e., backward into the period of actual 'Cellular Evolution'."[9]

To get there, Woese needed a genetic timekeeper, a way to classify microbes across vast evolutionary distances. He found one in a molecule called ribosomal RNA (rRNA), which synthesizes the amino acids that make a protein molecule in virtually all biological organisms.

The gene mutates slowly and passes from generation to generation with a high degree of transcriptional fidelity. Its near-universality makes it the perfect chronometer to trace the vertical descent of life over billions of years.

Together with his then-postdoctoral fellow George Fox and other graduate students, Woese set out to isolate and sequence one form of the molecule called 16S rRNA across a range of microbes. A research group led by Woese spent roughly a decade painstakingly pulling together a database showing the minute differences in 16S rRNA across a wide variety of microorganisms.

Creating a molecular atlas microbe by microbe required extraordinary concentration and persistence, yet by doing so Woese decoded the history of life and it would have enormous implications for twentieth-century biology.

Scarred Revolutionary

By the late 1970s, Woese started to realize that certain microbes, including those that produce methane and thrived in oxygen-starved habitats, were fundamentally different at the molecular level from the two existing domains of biological life: bacteria and the more complex eukaryotes that include plants, animals and us.

In 1977, Woese and Fox proposed that microorganisms called methanogens were actually part of an undiscovered third domain of life later known as Archaea.[10] It was an extraordinary claim that drew international press attention and met fierce resistance, even derision, from the biology establishment, including Salvador Luria, an Italian microbiologist and Nobel laureate of medicine for his work into how viruses replicate.

Another of Woese's adversaries turned out to be his long-time supporter: Ernst Mayr, an accomplished Harvard University evolutionary biologist and renowned taxonomist, who helped conceptualize the synthesis of genetics with natural selection, a body of work called Neo-Darwinism.

A classically-trained biologist wary of excessively mathematical approaches to evolution, Mayr argued that Woese's training as a physicist and unfamiliarity with the field's classification systems may have led him astray in proposing, erroneously in his view, his three-domain schematic for biological life.

"It must be remembered that Woese was not trained as a biologist and quite naturally doesn't not have an extensive familiarity with the principles of classification," Mayr wrote in a stinging criticism of Woese's work in 1990.[11]

The academic jousting over Woese's paradigm-shifting theory played out in pages of *Nature* and *Proceedings of the National Academy of Sciences of the United States of America*. Science writers portrayed Woese as a solitary, anti-establishment figure — "microbiology's scarred revolutionary," as one put it — who endured bullying by academic heavyweights in his dogged pursuit of his theory that prevailed in the end.[12]

There's some truth to those characterizations of Woese. Yet it takes nothing away from Woese's historic achievement to point out that such intellectual skepticism is the way that science is supposed to work. Woese wasn't just proclaiming that he had found an interesting new niche in life's microbial menagerie. His claim was more akin to the discovery of a vast new continent. It deserved scrutiny and received it in droves. One day that theory may well be overturned by a better one. Such is the way of science.

By the time he had died of pancreatic cancer in 2012, Woese's three-domain notion of biological life had become a generally accepted theory, enshrined in college textbooks. In the end, the guardians of Big Science showered him with his field's most coveted awards, from the National Medal of Science to the Leeuwenhoek Medal, microbiology's top accolade announced every 10 years. He never won a Nobel, but did receive the Royal Swedish Academy of Sciences co-sponsored Crafoord Prize for his work on the third domain of life.

Ironically, the discovery of Archaea, as consequential as it turned out to be, wasn't primarily what he set out to do. He developed a molecular and investigative approach to studying the speed and trajectory of biological life, wherever that might take us. The universality of 16S rRNA to catalog and sequence organisms could deliver massive insights about the evolutionary course of biological life, perhaps even its ultimate origins.

That was a staggering achievement in its own right. Yet his call to revise and expand our understanding of how the biosphere works also shaped the contours of his legacy. Woese believed that the massive amounts of genomic data coming on stream in the early decades of this century suggested that life might be just one facet of a bigger dynamic system that involved the fluxes of energy, chemicals and information, as well as genes and cells.

In the standard view, evolution reflects a process of continuous and gradual change across generations of life. Organisms and species were well-defined by their inherited genetic makeup. They would change but in a modulated way over time, from parent to offspring, via random mutations and the natural selection of better adapted organisms.

In complex and multicellular organisms, the large eukaryotes such as animals, that's generally the case. Yet in the microbial world, and probably during the early phases of evolutionary life on Earth, life was more chaotic and fluid, Woese argued. Studies showed that microbes traded, absorbed and discarded genes in real-time and across species in response to changing environmental pressures and attacks from predatory viruses.

In a 2007 essay titled "Biology's Next Revolution" published in *Nature*, Goldenfeld and Woese suggested that an avalanche of new genomic information about microbes in recent decades had challenged cherished notions about the nature of biological life and called for a reassessment of concepts such as organism, species and evolution.[13]

> *"It is becoming clear that microorganisms have a remarkable ability to reconstruct their genomes in the face of dire environmental stresses, and that in some cases their collective interactions with viruses may be crucial to this. In such a situation, how valid is the very concept of an organism in isolation? It seems that there is a continuity of energy flux and informational transfer from the genome up through cells, community, virosphere and environment. We would go so far as to suggest that a defining characteristic of life is the strong dependency on flux from the environment — be it of energy, chemicals, metabolites or genes."*

If a microorganism can radically change its genetic makeup during its lifespan, does it really belong to one species? Since genetic changes are occurring both horizontally across microbial species as well as vertically from parent to offspring, where does that leave our understanding of evolution?

In Woese's view, biology wasn't the study of being, but of becoming. Merely classifying organisms by shape, function and place in life's great pageant missed a far richer, more remarkable reality. The field needed to move beyond its fixation with taxonomy and broaden its intellectual ambitions, he argued.

Perhaps the tree of life, with its trunk and expanding branches, wasn't the right metaphor. Maybe biological life is more like a node in a larger and complex network, with cells and genes responding to changing states of energy and information that govern not just Earth but the physical universe.

Evolutionary Noise

Much as satellite-based astronomy had overthrown the conventional wisdom about the cosmos — researchers studying distant supernovae stunned the world in the late-1990s by announcing that the expansion of the universe was accelerating — so, too, enshrined ideas about organisms, species and evolution were being challenged by a vast amount of genomic data.

What we are learning is that the ways of microbes are very strange. Scientists were aware of the existence of horizontal gene transfer, or the non-genealogical exchange of genetic information from one organism to another, as far back as the late 1920s. Yet the arrival of new sequencing technologies and analysis of microbiomes showed just how pervasive such gene swapping really is in the microbial world.

It's now understood that gene transfers play a critical role in the worldwide expansion of drug-resistant microbes and the development of new pathogens.[14] Invasive bacteria such as *Vibrio cholerae* and *Streptococcus pneumoniae* spread toxins and exchange genes to stay a step ahead of antibiotic agents.

In larger colonies, microorganisms use chemical-signaling molecules in a process known as quorum sensing and their collective behavior is cooperative and strategic.[15] A family of bacteria called *Myxococcus* form thread-like networks to search for prey collectively. Other microbes communicate with each other chemically to determine if other microorganisms are friend or foe.

Then there's the role of viruses, as repositories and innovators of genetic information. Microbial viruses commandeer a host organism's cells to make copies and spread.

Organisms have been living with viruses for billions of years and the constant pressure has had an outsized impact on the cellular machinery of various species. Big data analysis of genomes suggests that the virosphere has played a bigger role in the evolution of life than other factors like predation and environmental shocks that impact food resources.[16]

In the early 2020s, scientists have far more sophisticated tools to work with, compared to zapping a flask full of chemicals with electricity à la the Urey-Miller experiment in the early 1950s. A research team led by Jack

Szostak, a Nobel-winning chemist and geneticist at Harvard University, is trying to find the pathways between nonliving chemicals and evolving, self-replicating life.

The group is designing and synthesizing primitive cells that existed billions of years ago. More specifically, that means recreating an ancient RNA-based nucleic acid genome and cell membranes that can grow and divide. The aim is to design a protocell that will then spontaneously make the leap to self-replication and evolving biological life.

A popular theory among origin-of-life researchers is the RNA World Hypothesis. In this scenario, RNA molecules predated both DNA and proteins as the basic parts of a cell. Billions of years ago, this ancient RNA stored genetic information, catalyzed chemical reactions, and developed the ability to convert inorganic raw materials into biological cells.

The idea that RNA played an outsized role in the origin of life was first floated as a possibility in the 1960s by British chemist Leslie Orgel, as well as Crick and Woese. Later on, as evolution developed more complex cellular structures, information storage became specialized in DNA and proteins handled the building and functioning of cellular life, according to this theory.

For years, not many paid attention to such musings since there wasn't conclusive evidence it was true. In those days, the only known biomolecules that scientists believed could trigger chemical reactions were proteins.

Then in the 1980s came the discovery, based on the work of molecular biologist Sidney Altman and chemist Thomas Cech, that a class of RNAs, called ribozymes, could catalyze chemicals and build proteins.

Their "discovery of catalytic RNA has altered the central dogma of the biosciences," in the words of the 1989 Nobel Chemistry award citation. "Moreover, it has already had a profound influence on our understanding of how life on Earth began and developed."[17]

Since then, the RNA World Hypothesis has become the organizing principle of a lot of origin-of-life research. "There must have been chemistry that drove the copying of RNA sequences," Szostak said in a 2019 lecture. "We think there were little bits of RNA floating around in a really rich chemical environment."[18]

Physics of Life

While Szostak and others are trying to create self-replicating life in their biochemistry labs, a theoretical biophysicist named Jeremy England has created a stir with a theory that the origin of life is the inevitable result of physics.

Back in 2013, while still at MIT, England came up with math equations behind his remarkable claim.[19] A group of atoms when faced with certain environmental pressures (such as solar and chemical energy, or the heat bath of primordial oceans) will restructure themselves in such a way as to capture energy and release heat more efficiently, according to his hypothesis.

Living things do that way better than non-living things. Under the right circumstances and given the physical laws of the universe as we know them, the emergence of life isn't all that surprising, in England's view. It's as inevitable as a rock rolling down a hill.

Looking at the origin of life through a physics lens is intriguing and England, who joined GlaxoSmithKline in 2019 as a senior director for artificial intelligence and machine learning, is perhaps one of the more creative thinkers approaching the problem from this perspective.

One of the first to do so was Erwin Schrodinger, the renowned Austrian-Irish theoretical physicist, in his book, *What Is Life? The Physical Aspect of the Living Cell*, published in 1944. In it, Schrodinger described life in terms of information and energy.[20]

Years before Crick and Watson characterized the double-helix, information storage role of DNA, Schrodinger portrayed chromosomes as "code script." He also suggested that life harnessed chemical energy in such a way as to maintain its orderliness in a physical system that trends toward disorder.

Schrodinger viewed biology through the prism of physics and, more specifically, thermodynamics. The second law of thermodynamics states that in a closed system, disorder always increases. Physicists call that disorder entropy.

Put another way, systems tend to disperse energy until they are evenly distributed. The classic example of this is the veritable cup of hot tea cooling off to room temperature. Air molecules leak from a balloon, but air molecules never return back into a balloon.

In a closed system, entropy (disorder) must increase over time. However, a living system can keep its entropy low by increasing the disorder of its surroundings. Living things grow and replicate by exchanging matter with their environment, and they're far better at capturing energy and releasing heat than non-living matter.

Physicists, including Woese, have made significant contributions to biology. They measure quantity, time, distance and other features of the physical universe, and then develop mathematically possible and predictive relationships between those numbers.

If you start with the working assumption that life consists of atoms and therefore obey established laws of physics, it's possible that life is an emergent consequence of these dynamics that we just haven't fully grasped yet.

England isn't your typical physicist. He's an ordained orthodox rabbi, who believes the bible has much to teach us not only about our yearning for life's most profound questions but also allegorically relates to the statistical thermodynamics that underpin his theory.

At the core of England's hypothesis is the notion of dissipative-powered adaptation, the notion that atoms tend to self-assemble into patterns to handle the energy flows of their environment. This energy harvesting behavior can take place with inanimate matter under the right circumstances.

"What is becoming increasingly clear is that interacting collectives of 'dumb' particles can evolve into specialized structures with fine-tuned relationships to their environment even in circumstances where there is no self-copying entity in the system to enable natural selection," England wrote in a 2020 essay.[21] "This dissipative adaptation is a broadly applicable physical mechanism based on the simple idea that energy both helps collections of particles to change their state of assembly, and is processed differently by a system depending on its current state of assembly."

England came up with equations based on established laws of physics that showed that atoms encountering the chaotic energy and chemical fluxes and heat conditions of early Earth would restructure themselves in such a way as to dissipate more energy.

"One feature common to all such examples of spontaneous 'self-replication' is their statistical irreversibility: clearly, it is much more likely that one bacterium should turn into two than that two should somehow spontaneously revert back into one," England wrote in his paper introducing his theory.

England's take on the origin of life is highly speculative and remains in the realm of an interesting idea. Nor has he produced a detailed explanation of how life actually started under his theory. However, he published a paper in 2017 in which computer simulations backed up his thesis about how life might have started.[22]

So the search goes on for the alpha of life's story. Microbes also may figure into humanity's destiny a billion years from now. Astrobiologists and theoretical physicists are coming up with some otherworldly ideas about seeding life in the universe. Ground control to Major *E. Coli*!

Chapter 15

Ground Control to Major *E. Coli*

Earth has an expiration date. As unpleasant as it is to contemplate, our planet won't be habitable forever. In the deep future, the sun will run out of hydrogen fuel in its core and start burning helium, becoming many orders of magnitude more luminous as it starts to expand and transform into a red giant star.

Astrophysicists believe that in about one billion years' time, the sun's increased brightness will cause a runaway greenhouse dynamic that causes atmospheric oxygen to plummet. Plants and organisms that rely on photosynthesis will die out, according to climate and biogeochemical models. The planet will no longer be livable for many forms of complex life.[1]

Microorganisms will likely outsurvive humanity for a time. Yet further out, in about five billion years, our dying sun will enter its red giant phase and start dramatically expanding. Eventually, it will destroy the inner solar system planets of Venus, Mars and Earth, according to the results of a stellar evolution study published in 2008.[2]

All of this assumes that biological life, or at least human civilization, will not be in serious trouble before then. In the centuries ahead, humanity could steadily destabilize the biosphere; nation states might trigger catastrophic nuclear wars; a single sociopath could design and unleash an unimaginably lethal viral super-strain; asteroid strikes, supervolcano eruptions or gamma-ray bursts could also result in a premature end.

Yet even if we somehow make it through the next billion years, our future descendants may need to find a new planetary home. The world's richest countries are investing billions into space exploration. They are contemplating missions to distant celestial bodies for reasons beyond just national prestige and the human yearning for answers to the big existential questions about our place in the universe.

Humanity needs a backup plan. If life becomes uninhabitable on Earth, our species will need to colonize other worlds or figure out a way to build sustainable, permanent habitats in space.

A lot of effort is also going into the scouting of exoplanets outside of our solar system that might maintain life, studying microbes to understand how we might genetically modify humans for long-duration space missions, and conceptualizing next-generation propulsion technologies needed for long interstellar journeys.

There are also decades of discovery science ahead to better understand the geological evolution of planets and what sort of life some of them can sustain. In our own cosmic neighborhood, the betting is that if life exists outside Earth, it will be microbial.

During the 1960s and 1970s, robotic planetary explorers that swept through our solar system found no traces of biological life. In 1976, America's Viking Lander conducted biological tests for microbial biosignatures in the soil of Mars, but the results were inconclusive.[3]

Since then, the search has continued for microorganisms on other planets and moons in our solar system. Powerful space telescopes, meanwhile, have discovered more than 4,500 exoplanets in other star systems, 300 of which may be hospitable to life.[4]

The range of potential future homes may increase in the early 2020s with the successful launch on Xmas Day of 2021 of the $10 billion James Webb Space Telescope, a collaboration among U.S., European and Canadian space programs, that will conduct origin-of-life research, among other tasks. The most ambitious robotic probe ever built, the Webb's infra-red technology, bigger mirror system and deeper orbit will allow it to peer farther back in time than NASA's Hubble telescope.

In recent decades, microbiologists have also learned far more about how microbes can survive in extreme temperature, acidic and radioactive environments and thrive without sunshine, oxygen or organic chemicals.

In 2020, Japanese astrobiologists conducting an experiment on top of the International Space Station, as part of the Tanpopo mission, announced that bacterial species survived three years of open exposure in space.[5] The scientists attached a capsule containing samples of *Deinococcus radiodurans* to the space station and observed that the microbes were able to withstand the high dose of ultraviolet radiation in space by repairing their own DNA.

Microbial Martians

In February, 2021, NASA's Perseverance rover, with a space helicopter attached, touched down on Mars to search for signs of ancient microbial life, as well as study the planet's geology and climate.[6]

The NASA mission landed Perseverance at the Jezero Crater on Mars that is believed to have been the site of a large lake and river delta system some 3.5 billion years ago. The robotic rover will be looking for microfossils within the 45-kilometer-wide (28-mile-wide) crater that might reveal biosignatures of microorganisms from the deep past.

The rover, equipped with an array of cameras, lasers and scanners, can analyze rocks for their chemical composition and retrieve the more promising ones with robotic arms for further analysis by NASA.

Searching the microbial fossil record on ancient rocks on Mars is in some ways easier than doing so on Earth. On our planet, the convection dynamics from our hot molten core creates a process of plate tectonics that constantly reshape the terrestrial landscape. The grinding and recycling of rock formations have wiped out parts of the historical fossil record.

In contrast, Mars formed as a searing mass of molten rock that cooled and formed a more stable crust around a rocky mantle. So there's a greater chance more of the fossil record, if one exists, to still be available for scientific observation. NASA scientists are on the lookout for rock formations called stromatolites, which are carbonate dome structures that on Earth contain evidence of microbial life.

In the second half of the 2020s, NASA plans to dispatch another lander to the polar region of Mars for astrobiological research. The Icebreaker Life mission will carry a drill to burrow into the frozen ground of the planet's northern plains to look for evidence of current or past microbial life.[7]

Paradoxically, the rush to explore Mars runs the risk of massive microbial contamination before all of the planet's secrets can be unlocked by scientists. Plans are underway for manned missions in the 2030s by government space programs and private aerospace companies such as SpaceX and Blue Origin.

Planetary protection policies like the United Nations Outer Space Treaty that require space-faring nations to prevent forward contamination of microbes from Earth may mean little once humans start exploring and living on Mars.[8] Some have called for a ban on manned missions to Mars until scientific studies of the Martian biosphere are completed.

Humans harbor trillions of microbes. Once an astronaut sets foot on Mars, a team of researchers argued in 2017, microbial contamination from Earthlings is a virtual certainty. "All this microbial diversity would potentially leak out of a spacecraft or habitat module or waste deposit, and some of the organisms could end up in Special Regions (for scientific research) because of transport by wind," they warned.[9]

Outside of Mars, another unlikely candidate for extraterrestrial microbial life has emerged. In 2020, scientists picked up a possible microbial biosignature in the acidic clouds of Venus, our hot, hellish and volcanic neighbor. An international research team detected traces of phosphine in the Venusian atmosphere using the James Clerk Maxwell Telescope in Hawaii.[10]

On Earth, phosphine is a gas produced by anaerobic bacteria that don't require oxygen for growth. Venus has never been anybody's idea of an ideal home for biological life, and some scientists are very skeptical of the findings, with a rival study suggesting the purported phosphine may actually be ordinary sulfur dioxide.

Thanks to its smothering, heat-containing atmosphere of carbon dioxide and sulfuric acid, Venus is the hottest planet in the solar system, despite the fact that Mercury is closer to the Sun. Surface temperatures are about 475 degrees Celsius (900 degrees Fahrenheit). Its landscape is covered with active volcanoes, while its air pressure is 90 times that of Earth.

The temperature within its clouds is far more temperate, but they're home to heavy concentrations of sulphuric acid. The possible existence of phosphine gas is a tantalizing theory but a very speculative one at this point.

At the end of this decade, NASA plans to launch the DAVINCI+ (as in the Deep Atmosphere Venus Investigation of Noble Gases, Chemistry, and Imaging) to explore the Venusian atmosphere to better understand how the planet evolved and whether it had an ocean.[11] Another mission, called VERITAS, will chart the planet's fiery topography and extensive volcano system.

Astrobiologists have extended their search for alien microbial life beyond planetary environments. Rich in organic material and water as they are, comets may be home to extremophile microorganisms. Theorists have long speculated that microbial life may have been transported from one planet to another via space dust, meteoroids, asteroids and comets.

As far back as 1871, in an address to the then-called British Society for the Advancement of Science, mathematician and physicist William Thomson, better known as Lord Kelvin, suggested that meteorites could have propagated the spread of life.

The so-called panspermia hypothesis is a tantalizing one, but has never been fully embraced by the scientific community. Still, within our solar system, there's little doubt that Earth has exchanged space rocks with Mars.

NASA caused a global sensation when its scientists announced in 1996 that a fragment of a Martian meteorite found in the Allan Hills in Antarctica back in the mid-1980s showed tentative evidence of microscopic fossils of bacteria.[12] However, the finding was met with a wave of skepticism by other scientists, who believed the analysis was inconclusive.

In recent decades, researchers have come around to the idea that microbes could survive in space for years, despite the exposure to heavy ultraviolet radiation. Astrobiologists have shown that some bacteria have a survival mechanism in which a colony surrounds itself with a biofilm that greatly increases protection against UV and other harmful radiation.

Other microbes can go into a dormant phase to survive harsh space travel and reawaken later. That said, despite decades of searching, Earth remains the only place that life is known to exist.

Europa Beckons

Astrobiologists have widened their search for microbial life to Europa, an icy moon of Jupiter. NASA is planning to send a spacecraft, as part of the Europa Clipper mission, to explore the orbiting Jovian celestial body in 2024.[13]

NASA's Galileo spacecraft orbited Jupiter from 1995 to 2003 and based on readings of the moon's magnetic field surmised the possible existence of a global ocean of salty water underneath an icy surface. In fact, scientists believe there's possibly more water on Europa than exists on Earth.

The Europa Clipper will take numerous flybys of the moon surface later in the decade. Researchers believe there may be geological activity and heat generation on the floor of the ocean from a process called tidal flexing. Europa gets stretched and released by the enormous gravitational pull of Jupiter, which is more than twice as massive as all of the other planets in the solar system combined.

The probe will be looking for signs of the essential chemical ingredients for life: carbon, hydrogen, nitrogen, oxygen, phosphorus and sulfur. There's a reasonable chance that they were part of the formation of the moon or deposited later by comets and asteroids.

If life on Europa does indeed exist, it will most likely be found in the Jovian moon's subterranean waters. Jupiter emits massive amounts of radiation on its moons, which is bad for life on the surface but a potential energy source for underground oceans. Should the existence of a salty ocean, diversified chemical elements and hydrothermal vents be confirmed by the Europa Clipper, it's possible that there's bacterial life.

Microbes, meanwhile, are far better designed for long-term space travel than complex life forms like humans. Beyond our planet's protective magnetosphere, biological life in deep space is bombarded by high-energy ionizing radiation that damages DNA and raises cancer risks.

Over a long period of time, a reduced gravity environment causes bone density loss, muscle atrophy and cell damage. Bacteria in space environments can turn more virulent or antibiotic resistant.

Various space programs and the International Space Station have conducted microbiology research to see how bacteria like *Escherichia coli* and *Bacillus subtilis* adapt to heavy radiation and microgravity states.

Since the mid-2000s, American and European space agencies have conducted microbial experiments using tiny space vehicles called CubeSats equipped with biosensors to track cell growth and gene expression in harsh environmental conditions in space.

These cube-shaped research spacecraft, also called nanosatellites, are typically built to dimensions of 10 cm × 10 cm × 10 cm and weigh about 1.33 kg (3 pounds) and are flown into space as auxiliary payloads on scheduled space missions.

The American Artemis-1 mission, an unmanned deep space exploration system scheduled to launch in early 2022, is expected to carry a BioSentinel Satellite, which will transport genetically modified strains of the yeast *Saccharomyces cerevisiae* to see how they react to long-term, deep space travel.[14]

The yeast's DNA repair system is similar to those of humans. Scientists hope to gather data that will help reduce the risk that comes with long-term human space journeys. BioSentinel is the first nanosatellite to conduct biological experiments in deep interplanetary space.

Terraforming Mars

Within our solar system, most scientists believe that our best shot for a viable planetary colony is Mars. Yet not in its current state: Low average temperatures, scant oxygen, punishing ultraviolet radiation levels and the lack of liquid water pretty much rule out the survival of human life without a habitat.

The idea of geologically engineering, or terraforming, planets has been the stuff of science fiction novels since the 1940s. However, scientists, including cosmologist and astrobiologist Carl Sagan, started considering the issue in a serious way more than 60 years ago.

In 1961, Sagan suggested in a published paper that it might be possible to alter the atmosphere of Venus with genetically programmed bacteria designed to eat atmospheric carbon and produce organic molecules as a first step toward refashioning the planet's environment. [15] A decade later, Sagan proposed that vaporizing the northern polar ice cap of Mars would create an atmosphere over the planet, raise temperatures and increase the likelihood of liquid water.[16]

Tesla founder and technologist Elon Musk with his trademark brashness suggested in 2015 that humanity should consider dropping nuclear bombs on the planet's icy polar regions.[17] By his reasoning, the explosions would release enough water vapor and carbon dioxide to create a greenhouse effect and warm up the planet.

In 2019, a group of scientists argued for a radical departure from existing international government space policies that strive to prevent microbial contamination between planets. Instead, microbes should be used as our first wave of Earthly life forms to colonize new worlds.

"Assuming that a colonization plan aims for eventual permanence, the first colonists should consist of microbial species, not human, paralleling what likely happened on primordial Earth," they argued. "The paradigm shift we now advocate is that a deliberate seeding of microbes would ultimately promote colonization goals — e.g. terraforming."[18]

Terraforming Mars, turning an airless, dry and frigid planet into an Earth-like habitat, would be the most audacious engineering project ever attempted by humanity. It would require artificially melting the planet's polar caps to produce enough liquid water and carbon dioxide to increase atmospheric pressure and temperature to protect life from radiation.

In short, we'd need to create a type of runaway greenhouse effect on Mars, creating a feedback loop in which rising temperatures heat up the planetary surface, which in turn releases even more CO_2 into the new atmosphere.

Even if we developed the processes to do this effectively and responsibly, some scientists doubt there's enough raw material on Mars to pull it off. In 2018, two NASA-funded researchers did the math and concluded there wasn't enough CO_2 on Mars to trigger meaningful greenhouse warming. Using all available resources would increase atmospheric pressure to only 7 percent to that of Earth.[19]

Transforming Mars might be doable on a regional basis by coating the surface with a translucent material like silica aerogel that would create water and heat in select areas, according to Harvard researchers.[20] However, any future human Mars colony would most likely need to live in vast habitation domes to cultivate crops and to survive.

To move permanently into space, we will need to develop industrial technologies such as biomining to better extract raw materials from colonized planets and asteroids.[21] Some microorganisms can basically eat through stone, and have been useful on Earth in helping release elements such as gold and copper from rocks.

European scientists performed what they called the European Space Agency BioRock experiment on the International Space Station to see how biological mining of basalt, a material ubiquitous on Mars and the Moon, worked in extraterrestrial gravity conditions. The microorganisms under study such as *Sphingomonas desiccabilis* and *Bacillus subtilis* had little trouble handling the shift in setting, according to the results of the study.

Interstellar Visions

Even if we manage to find a new planetary home within our solar system, that won't help us much when our sun dies off in the deep future. The ultimate insurance policy is to find a habitable exoplanet in another and younger star system.

However, that scenario poses formidable aerospace engineering challenges. The closest star to Earth is Proxima Centauri. That's about 4.25 light years away, as in 4.25 years of travel at the *speed* of light. Our current propulsion technologies are inadequate for a journey of that distance.

We've come a long since Wilbur and Orville Wright achieved the first successful airflight with the Kitty Hawk in 1903. Yet despite the remarkable achievements of space programs since then, only a handful of spacecraft have actually left the solar system: the Voyager 1 and Voyager 2 probes launched in 1977. Lifting massive objects into space at the escape velocity needed for interstellar travel requires a type of breakthrough rocket technology we currently don't have.

One way around this might be to launch very small spacecraft powered into space with directed energy sources like lasers. Scientists and engineers believe ultra-thin "wafer sats" with miniaturized sensors and communication circuits, weighing no more than a gram, could be hurled into space at speeds of 25 percent to 30 percent the speed of light.

Originally conceived by University of California, Santa Barbara physicist Philip Lubin, the Directed Energy Propulsion for Interstellar Exploration is a spaceflight concept now under development by NASA that envisions an array of lasers that would be synchronized and pointed at a reflector, or laser sail, attached to a miniature spacecraft.[22] The laser technology development work is similar to ongoing research into a planetary defense system against asteroids.

"That would enable us to get to the nearest star, Proxima Centauri, which is a little over 4 light years away, in less than 20 years," according to Lubin.[23] The lasers would only be needed to achieve escape trajectory into space. "For small spacecraft, it's only for a few minutes, and then it's like shooting a gun. You have a projectile which just moves ballistically."

We are probably more than three decades away from laser-powered space travel, according to Lubin. One big challenge is decelerating the tiny spacecraft as it approaches its target planet.

Yet if realized, directed energy propulsion technology could have many valuable applications. Space travel within the solar system could be greatly accelerated. Supplies could be ferried to Mars in just a few days. Power and communication could be transmitted to spacecraft and bases on the moon. Threatening space rocks could be diverted from Earth's path.

All of this in theory opens the door to scientists sending microbes and nano-machines to neighboring star systems to seed life and build infrastructure for an interplanetary human civilization in the centuries ahead.

Harvard biologist, geneticist and futurist George Church is working on a speculative concept that involves spacecraft at the scale of a nanogram, which is one billionth of a gram. The basic idea is to send microorganisms and biological cells that could replicate but also be genetically designed to perform tasks.

"There are challenges, definitely, but a nanogram is about the size of a eukaryotic cell. We know that a eukaryotic cell, a single cell, can contain enough information to create a very complex body," Church has explained.[24] "It could be programmed with enough information to build a light source that could beam back, bidirectionally, establish a communication line. Then you could start moving things at the speed of light. Not matter, but information."

Genesis Project

One of the more remarkable concepts of a future interstellar age comes from German physicist and novelist Claudius Gros. A professor at Goethe University in Frankfurt, Gros has studied the feasibility of long-distance space travel and published a paper in 2016 proposing his Genesis project.[25]

Gros thinks the number of potentially habitable exoplanets is vast, though they will come in endless varieties. Much will depend on the composition and age of the host star and the mineral composition, water volume and atmosphere of the target planet for exploration and colonization.

"In the universe, current estimates are that you probably have billions of potentially habitable planets," Gros told me. "Most scientists believe that microbial biological life is pretty common."

If biological life exists on other planets, Gros thinks the odds are high that it will be in evolutionary stages similar to the early, single-cell microorganisms rather than more complex life forms, which took about two billion years to emerge on Earth. There are also perfectly habitable worlds that for whatever reason are biologically barren.

Gros' research interests range from computational neuroscience to self-organizing robots. He became interested in humanity's deep future after visiting California years ago and examining the geological markings of petrified wood a couple of hundred million years old.

That glimpse into the past made him think about the distant future, and how biological life might survive a billion years from now. Gros is quite confident that laser-powered space travel of wafer-thin spacecraft will be a reality 50 to 100 years from now.

"It's very, very difficult, but it's normal technology," Gros told me. "It's not Star Trek." He thinks the biggest engineering challenge will be figuring out how to slow down high-speed spacecraft as they approach the orbit of their final destination in distant space. "The key issue is that it is much more difficult to slow down than to accelerate."

In his conceptual framework, Gros imagines an armada of tiny, but technologically sophisticated, spacecraft approaching an exoplanet and scanning for biological life.

If complex life exists, nothing happens. If the planet is barren, an onboard genetics lab will synthesize a mix of unicellular and more complex microbial life calibrated to local conditions and send them down to the planet's surface with the aim of kick-starting an evolutionary dynamic.

Gros expects a messy, chaotic process in the early phases of this man-made genesis of life project. "In the first few thousands of years there will be one ecological disaster after another. There will be no stable ecosystem at the start," he figures. "In a few hundred thousand years, you'll get a more stable ecosystem."

The emergence of complex life may take hundreds of millions, or even billions of years, to take hold. For Gros, the Genesis project has little to do with humans. The aim here is to seed life around the universe, not necessarily human life.

For one thing, the vast time frames required mean the generation of humans who pull off the first genesis experiment won't be around to see the results. Secondly, evolution is inherently random. There's no guarantee that complex life forms that evolve on an exoplanet will resemble modern humans. "Evolution isn't directed. You can't give it an aim. The end product will be interesting," Gros muses.

That said, if humanity itself wants to thrive on exoplanets in the deep future, Gros thinks it will require a fleet of robotic spacecraft in several waves. The first will scout out the planets and then spend thousands of years building a human habitat. Raw materials can be extracted from space rocks in the vicinity.

The second robotic team would build a biotech lab to genetically design humans to be grown on the alien planet. That kind of know-how should be available a century or so from now.

Gros has thought quite a bit about the ethics of all of this. He thinks that planets with existing life forms should be left alone, yet he allows that there's a risk of an accident that destroys foreign life forms.

At the same time, Gros believes that it's not useful to apply environmental protection ethics that make sense on Earth on exploration of the wider universe. In a closed system like our planet, preserving a sustainable environment and biodiversity is literally a matter of life and death for our species.

Alternatively, if alien microbial life is unintentionally destroyed by human exploration on a handful of planets, the entire existence of the universe isn't at stake. "Of course, we don't want to destroy life at all. But to generalize the protections we have on Earth to the galaxy, I find it difficult to find a rationale for that," Gros reasoned.

If Gros' strangely alluring vision of the future is realized, humanity will have to get its mind around a cruel irony. After our sun dies and the Earth is no more, life on this planet may yet find a new home elsewhere. It just may not be human life. Microbes preceded us and will most likely outsurvive us.

Conclusion: What the Microbes Are Telling Us

In the popular imagination, scientific discovery is an inexorably rising arc of knowledge leading humanity to the sunny uplands of something we call progress. Thomas Kuhn, one of the most influential philosophers of science, suggested that how we figure things out about the world around us is a far messier, more nonlinear process.

Kuhn, in his 1962 book, *The Structure of Scientific Revolutions,* identified two types of knowledge acquisition that have emerged since humanity embraced the fabled scientific method: systematic observation and precise measurement combined with the formulation, testing and refinement of hypotheses.[1]

In Kuhn's telling, there are long stretches of what he calls normal science, incremental gains in knowledge based on established theories that expand our perception of the natural world. "Normal science," Kuhn wrote, "means research firmly based upon one or more past scientific achievements, achievements that some particular scientific community acknowledges for a time as supplying the foundation for its further practice."

Then, sporadically, that foundation of knowledge gets shattered. There's a sudden disruption in our sense of things, a revolutionary scientific discovery overturns established truths, forever changing the conversation and research priorities. Conservative guardians of the older, discredited theories may counter-attack for a time, but eventually a new consensus takes form.

"Paradigm changes do cause scientists to see the world of their research engagements differently," wrote Kuhn. "Insofar as their only recourse to that world is through what they see and do, we may want to say that after a revolution scientists are responding to a different world."

Does the amazing run of microbial scientific breakthroughs in the first two decades of this century rise to the threshold of revolutionary discovery that Kuhn envisioned? One of the twentieth century's most acclaimed biologists believed so.

As we discovered earlier, Carl Woese, credited with discovering a new domain of microorganisms — the Archaea — and upending established dogma of his field in the late-1970s, believed microbes challenged our current understanding of what it means to be a biological species and evolution.

Woese trained as a physicist before making his reputation in biology at the University of Illinois, Urbana-Champaign. He saw parallels between the stunning leaps in observational cosmology during the 1990s and the equally, mind-altering realization of the size, scope and pervasiveness of microbial systems in our bodies, in the biosphere, and in the biogeochemical processes that run our planet.

"We now know that what astronomers used to think of as 'the Universe,' the visible universe, is less than 4% of the total matter/energy density of the universe, the rest being made up of the still-mysterious dark matter and energy," wrote Woese and Nigel Goldenfeld in a 2009 essay.[2] "Similarly, we now understand that in terms of both numbers and genetic diversity, the microbial world not only dominates the biosphere but is almost impossible to sample properly."

Even if you accept that microbes are fascinating, worthwhile targets of scientific inquiry, there's still the issue of why we human "macrobes" should care. With all due respect to the molecular wonders of *Pseudomonas syringae* and *Escherichia coli*, one might ask, why should we spend much time thinking of these imperceptible creatures at all?

Humans tend to have a myopic view of life, given how accustomed we are to viewing living things on our visual scale. Individual bacteria and viruses are invisible to us. Unless they directly threaten our health, they don't appear much on our attention radar screens.

Situational Awareness

One of the aspirations of this book has been to convince you to consider otherwise. This project was conceived, researched and written in the middle of a global health emergency, courtesy of a super-transmissible, mutating zoonotic coronavirus. It has killed approximately six million worldwide as of the spring of 2022, and destroyed livelihoods. The long-tail health impact of COVID-19 means millions more will be living with its lingering after-effects for years to come.

We have only a limited understanding of the future pathogenic threats to our plants, livestock and human populations, while antibiotic-resistant bacteria and fungi are on the rise in an interconnected and warming world. Microbial systems are intrinsically tied to this century's most complicated problems.

Population pressures and heavy chemical fertilizer use stress our terrestrial resources, weaken soil health and erode topsoil. Our food security depends on a better understanding of the terrestrial microbiome that controls nutrients and nitrogen recycling. In our increasingly acidic and warming oceans, shifting phytoplankton populations may be influencing food webs and carbon dioxide absorption.

The contours of climate change in the century ahead will be shaped by methane-producing microorganisms from the Arctic to the Amazon basin. Widely dispersed dual-use technologies that allow us to design and synthesize DNA, genetically modify microorganisms, edit genomes, and even create entirely new biological entities pose arguably far more worrisome proliferation risks than the nuclear age has thus far.

There are also, of course, tremendous potential upsides for human health, environmental remediation, and our quest to decarbonize the economy, if we can harness microbial biochemical and catalytic enzyme superpowers. While disease-spawning viruses and bacteria rightly receive urgent attention, it's worth remembering that most microorganisms are actually beneficial to humanity or entirely harmless.

With more breakthroughs and research at the world's universities, industrial biotech and synthetic biology labs, the biochemical properties and proteins being teased out of the microbial dominion may improve the efficacy of our cancer drugs, Alzheimer's treatments, antibiotics and vaccines. The swirling microbiota residing in our bodies could be managed in a beneficial way to fine-tune our immune systems and manage the neurochemicals that shape our moods, and improve our subjective experience of living.

As we decarbonize the economy, biological materials perhaps will help displace plastics and fossil fuels that threaten the environment. What we eat, how we farm, what materials we use to construct our homes and cities, our physical world — they all could be transformed in the decades ahead powered by the biotransformation of our economies and societies.

The rate of these changes likely will run at varying speeds in different parts of the global economy. Some industries will be slower to make the transition than others. Microbial science married to digital precision farming is already starting to transform Big Agriculture in the U.S., Europe, Russia, Ukraine and China.

In the developed world, "if you look at what's happened in the last five years with digital farm management systems driving decision-making in very, very complex tradeoffs, there's been an unbelievable, exponential uptake," explained Martin Clough, Head of Technology and Digital Integration at Syngenta based in Switzerland.

The industry-wide embrace of biofertilizers and gene-edited microbial solutions will very much depend on economic incentives. "Our job is to gain scientific insight, then make it really easy to adopt and put into practice. Then make it financially attractive to actually do it," Clough told me. "I think that before the end of this decade you'll see a substantial change."

Ester Baiget, the President and Chief Executive Officer of the Danish industrial biotech company Novozymes, believes that biology, microbial science and biotech are now certainly advanced enough to help societies transition away from fossil-fuel powered economies.

"We live in a very special time," she explained when I visited her at the company's headquarters just outside of Copenhagen. "Yes, we are in the midst of a pandemic, but we also have to embrace that to continue doing what we are doing is simply not sustainable."

How quickly that coming transformation will happen will turn on two dynamics in her view. The first is how much climate stress the world is willing to accept before making a decisive break from oil, gas and coal as primary energy sources.

Fossil fuels are woven into the fabric of the global economy and the energy industry is still big, financially powerful and politically influential. The second is the ability of biotech companies to take promising ideas from prototypes in test tubes to commercially viable products distributed across global markets.

"To make biological products that are competitive to fossil-based ones, you not only need to design them, you need to bring them to scale. This is the bottleneck today," according to Baiget. "You have five grams of something that will be fantastic. Great. But how do I make it across the globe, make it reliable, safe, and in a way our customers can utilize," she said.

Microbial Indifference

It is difficult to predict which alternative future is in store for us. Will there be endless disease outbreak cycles and accelerated climate change or extended human lifespans and sustainable, circular economies? Elements of both scenarios will likely shape the human story in the decades ahead.

Microbes are completely indifferent to us. They react to evolutionary pressures and like any other biological entity seek to grow, survive, and reproduce. Our relationship is adversarial, symbiotic and transactional. It obviously isn't personal. Humanity and microbes are two dominant life forms in the biosphere. As such, our fates are intertwined.

Lens out a bit more, though, and it's clear the biosphere that we share is but one node in a far more intricately, complex Earth system network. It's one shaped by physical and energy fluxes, as well as feedback loops that flow throughout the atmosphere (air), hydrosphere (water), cryosphere (ice), geosphere (rocks) and lithosphere (Earth's crust).

Many of these processes operate on different time scales from the evolution of human life, winding back to the early moments of the Earth's formation 4.5 billion years ago. Life occupies a tiny, precarious physical space in the Earth supersystem.

"The biosphere, all organisms combined, makes up only about one part in 10 billion of the Earth's mass. It is sparsely distributed through a kilometer-thick layer of soil, water, and air stretched over a half billion square kilometers of surface," the American biologist E. O. Wilson wrote in his book *The Diversity of Life.*[3]

"If the world were the size of an ordinary desktop globe and its surface were viewed edgewise an arm's length away, no trace of the biosphere could be seen with the naked eye," added Wilson, who passed away at the end of 2021. "Yet life has divided into millions of species, the fundamental units, each playing a unique role in relation to the whole."

Intertwined Fates

So why should microbes warrant our attention? For starters, they are highly adaptive and have thrived for nearly four billion years. They generate much of the planet's oxygen supply, decompose our dead organic matter and waste, recycle atmospheric gases, and produce nutrients, vitamins and biochemicals that keep us alive and the biosphere humming.

We're houseguests in a microbial world. Microorganisms represent two of the three domains of life (Bacteria and Archaea), as well as many of the kingdoms of the third one, Eurkarya. They are the irreplaceable operating system for a global ecosystem that supports all life forms.

Since we can't live without them — our ocean health and food webs would collapse if they vanished — it's a safe assumption they will outlive us in the deep future. They're better suited to interstellar space travel than humans and may end up being our best bet to seed biological life from Earth on habitable exoplanets.

We need a far better scientific awareness of how these microbial systems work, the threats they pose, and the benefits they offer. In researching and reporting out this exploration of all things microbial, it's clear the cost of ignoring how our actions reshape the microbial universe could be very steep indeed.

Three urgent tasks stand before us in the 2020s. We need to invest in artificial intelligence-powered pandemic preparedness systems, fashion a new biosecurity framework designed for the age of synthetic biology, and, finally, better grasp how microbiome networks impact agriculture, ocean health and climate change.

Pandemic Preparedness

More than two years into a pandemic, the world now has a test case of the destructive impact of a highly transmissible respiratory virus. Yet the COVID-19 pandemic wasn't the *big one*. Not even close.

As we've seen, epidemic risk firms like Metabiota that run models of hypothetical respiratory viral or influenza outbreaks see potentially far bigger catastrophic events on the horizon. With the right characteristics, a new virus or existing influenza could inflict damage never seen before in human history. In the 20 worst-case, multi-year pandemic scenarios simulated by Metabiota's models, the average number of expected deaths were 264 million globally, ranging from 117 million to as high as 545 million.

What's more, the next global contagion that nature produces could be more lethal and far more molecularly complex than SARS-CoV-2. We knew quite a bit about coronaviruses. Other viral families we know far less about. Remember that scientists have spent nearly four decades trying to develop a vaccine for HIV.

Infectious diseases don't arrive on a fixed schedule, and once a crisis fades from a generation's memory the world has a tendency to move on. However, the tempo of outbreaks during the first two decades of this century has increased in frequency. As human populations, agricultural lands and urban sprawls collide with dwindling natural habitats, the probability rises for pathogens to spill over from animal reservoirs to urban centers.

The U.S., Germany and the World Health Organization are now starting to map out investments in early warning systems. Yet the scale of the investment has to meet the challenge at hand: microbial disease threats are heterogeneous, complex and fast-moving phenomena, requiring a data and analytics platform that can deliver timely and reliable information from all over the world to multiple actors.

Governments build intelligence satellites to track emerging threats. Similarly, we need that kind of technological sophistication and investment at the molecular level, cataloging all manner of biological information — circulating viruses, bacteria, metabolites, antibiotic-resistant genes, as well as genetic sequences of microorganisms in terrestrial, oceanic and atmospheric environments. The more data that inform machine learning programs, the smarter the AI.

Building a multi-layered pandemic prevention system to detect, assess and respond to new disease threats will require $357 billion worldwide over a decade-long period, according to a study by McKinsey & Company. Sounds steep, and it is. That still works out to roughly an average cost of $4 per person every year.[4] And remember: The total financial hit from the COVID-19 pandemic from 2020 to 2025 will reach $22 trillion by one estimate.[5]

Smart systems that use AI programs to identify the protein structures and molecular makeup of dangerous microbes will be crucial for timely vaccine and treatment candidates. Disease outbreaks are societal problems requiring a societal response. Relying on a handful of government and public health elites to save the day is a recipe for failure.

Corporate risk-managers, point-of-care healthcare workers, educators, urban leaders, and ordinary citizens need to be educated and primed to play a role in pandemic response. Disinformation campaigns by state actors and social media networks spreading false public health information have to be chased down, sanctioned and discredited. We'll need to war game emerging disease threats with the same intensity we manage national security risk. For most countries, a pandemic that kills millions is actually a far higher probability than a land invasion by some foreign adversary.

None of this will be easy or even offer full-proof protection against future pandemics. It will, however, allow us to move in a swifter and more concerted way to contain and manage epidemics before they destabilize the world again.

We've tried managing a worldwide disease outbreak on the fly, with little or no protection. It hasn't gone well. If long-term global stability is a goal of the human species, then managing pandemic risk is a necessary, not optional, investment.

Biosecurity

We have a Cold War-era bioterrorism defense strategy dangerously out of sync with the awesome capabilities of today's biotech and synthetic biology industries. Multiple actors in industry, academic research and even amateur biohackers have access to powerful tools that can design cells, synthesize DNA and rewrite the code of life.

Weaponizing a virus, toxin or other biological agents is technically difficult, but counter-terrorism experts see serious vulnerabilities in the world's fragmented and uneven biosecurity system that might be tempting to a state actor or a rogue scientist seeking to do great societal harm.

Life science companies, particularly in the synthetic biology field, have invested in biological risk assessment screening and scrutinize orders involving sensitive research. Yet the U.S. Government guidelines they follow are more than a decade old and rely on restricted lists of pathogens that are outdated in a biotech industry that can genetically modify, or even basically create, new microbes.

A clever bioterrorist could evade these restrictions by ordering up synthesized microbes that don't map to government pathogen lists, or strategically engineer smaller, individual genetic sequences that when stitched together are far more menacing. One could triage DNA production companies around the world for a synthesizer provider that doesn't ask too many questions, or will skip a biosecurity screening altogether for the right price.

Companies are developing mobile gene sequencers and DNA synthesizers that would allow researchers to do high-powered biology in all sorts of settings hidden from public view. The full genetic sequence of smallpox and other pathogens can be downloaded from the Internet anytime, anywhere.

Biosecurity labs that do sensitive work with dangerous viruses and bacteria aren't well regulated in many parts of the world. One reason the unsubstantiated Wuhan lab leak theory of the COVID-19 pandemic seemed so credible is the long history of accidents at the world's high-security biolabs that work with infectious agents. This needs to be tightened up, given the upswing in vaccine and microbial pathogen research that will be taking place in the wake of the COVID-19 pandemic.

Microbiome Research

Microbes are amazingly collaborative life forms. They communicate via chemical signals called quorum sensing, swap genes between species and shape-shift their interactions as they respond to new information in their environment.

Improving our scientific knowledge and situational awareness of microbiomes — vast networks of microbial species in the body, land, oceans and atmosphere — will require years of more basic research.

Research money is flowing into the fledgling field of the human microbiome, considered to be our second genome by some. With advances in engineering, protein structure analysis, genetic modification and gene editing, agricultural seeds and biofertilizers and other microbiota products are one of the fastest growing sectors in agronomy.

Microbiome science could also play a future role in greenhouse gas recycling, storage and geoengineering projects to prevent a climate breakdown from centuries of man-made, fossil-fuel emissions.

Talk to the experts in microbiome research, and it's clear that there are some big research obstacles standing in the way. There's a lack of data standardization. Scaling up research to go deeper into microbial swarms in aquatic and stratospheric environments is expensive.

Consider that the human microbiome consists of billions of strains of bacteria, fungi, viruses and other microorganisms. It's not hard to take a sample of this cacophony of microscopic life, but figuring out which organisms do what and how they all interact is extremely difficult.

Data are often contradictory and mapping specific microbes to function is still a very imprecise science. Human microbiome studies have found interesting correlations between species of microbes and health. There has been far less success pinpointing the exact causal relationships between a microbial gene, strain or microbiota community to a specific disease process.

We are some ways away from developing treatments that cover the whole spectrum of behaviors in a diverse microbiome. The same holds true in agricultural microbial science. Dreams of nudging soil microbes to absorb more of a greenhouse gas like nitrous oxide from the atmosphere as part of their nitrogen fixing role with plants will take years of additional basic research.

Perhaps the most glaring gap in our knowledge is how the undulating realm of microbiota in the biosphere is responding to climate change and what that means for carbon dioxide, methane and nitrous oxide emissions.

Microbes play an outsized role in the feedback loops of global warming, but aren't adequately reflected in the supercomputer-powered climate change models that inform policymakers about temperature trends and greenhouse emission targets.

Throughout Earth's geological history, microorganisms have shaped our atmosphere, starting with the Great Oxygenation Event some 2.4 billion years ago as photosynthetic bacteria in our seas started to generate O_2 in significant amounts. Microbes both produce and consume carbon dioxide and methane as they recycle chemical elements and consume energy.

We don't have a good sense of how a warming climate is impacting the behavior of these microbes as the permafrost zones thaw and free up organic material stored away for millennia. Both the Arctic and the Amazon Basin have been carbon storage systems for the planet. There's some evidence that human activity and global warming are already turning these regions into net emitters of greenhouse gas, or will so later in the century.

If vast amounts of carbon sequestered in these regions are released into the atmosphere, our climate risk could be vastly more concerning later in the century. We won't know that until we better understand how these microbiome recycling systems work and they're properly reflected in our climate change computer models. It would be species malpractice to do otherwise.

Evolutionary Mindshift

More insights into the microbial world are urgently needed, yet what's been learned thus far in the last two decades has already shifted perceptions of what it means to be human, a species, and evolutionary change. Microbes blur those distinctions. How human are we if we're home to roughly as many microbial cells as human ones?

If biological life is a swirl of interconnected systems that transcend species, what do we miss by not considering the existence of metaorganisms: that is, specific life forms and their associated microbiota. If both are an inseparable functional unit, that would have theoretical implications for any number of scientific disciplines and industries.

Back in 1991, the evolutionary theorist Lynn Margulis popularized the concept of a "holobiont," a term that refers to a biological entity that consists of a host and its symbionts.[6] Some microbiome scientists have embraced the concept on a broader scale to describe the interaction between humans, plants and animals, as well as networks of microorganisms.

Scientists in this field also talk of the hologenome to describe a host and its associated microbial gene pools. A human hologenome, for instance, consists of a human's roughly 20,000 genes and the estimated 33 million genes of the typical body's associated microbiota.

There's a debate among evolutionary biologists about whether the holobiont, and its associated hologenome, should be considered as a single unit of natural selection. In other words, should the evolutionary fitness of a species take into account the fitness of its microbiome?

Whether species and their microbiomes co-evolve in nature is an intriguing idea, but extremely speculative at this point. What is clear is that as we explore new drugs and treatments, aim to improve ocean health and food security, and chart the future climate change of the planet, microbial behavior will play a significant role whether we acknowledge it or not. This unseen majority has a seat at the table.

What happens in the microbiome, doesn't stay in the microbiome. In a closed, interconnected supersystem like Earth, one flutter of the veritable butterfly's wing can have cascading secondary impacts that ripple through the wider network.

In the world of microbiomes, environmental shocks to their systems can have implications for their host. Antibiotics destroy targeted bacteria, but also many other microbes inside the human body, often causing unpleasant side effects and opening up patients to other opportunistic microbial infections. In the broader context of the Earth's multifaceted systems and the long-term future of the human species, changes in the microbial system can have systemic spillover effects as well.

The microbiota running the planet are incredibly resilient and can absorb new stresses for quite some time. Yet complex systems do experience tipping points that are very difficult to reverse once the threshold of a phase transition has been breached.

Reawakened methane-producing microbes in the fast-warming Arctic could overwhelm the populations of other microbes that consume the gas. Vast amounts of methane could be consequently released into the atmosphere. Massive algae blooms, triggered by nitrogen runoff from agricultural lands, conceivably may create even bigger oxygen-dead zones in our oceans, devastating fishing stocks. What makes evolutionary sense for microbes in the decades ahead could turn out to be extremely harmful to human civilization.

Stephen Jay Gould, the American paleontologist and evolutionary biologist, marveled at the "unparalleled variety" and "unmatched modes of metabolism" of bacteria.[7] "Our shenanigans, nuclear and otherwise, might easily lead to our own destruction in the foreseeable future," he wrote. "We might take most of the large terrestrial vertebrates with us — a few thousand species at most. I doubt that we could ever substantially touch bacterial diversity."

So it might be wise to listen to what the microbes are telling us. They are the biosentinels of planetary health and will shape the contours of human destiny. They know a thing or two about what it takes to be a successful life form. And they're in it for the long haul, even if we're not.

Acknowledgments

Bizarrely enough, my interest in the microbial world owes a small debt to the Japanese terrorist Shoko Asahara. His doomsday cult, Aum Shinrikyo, launched a brazen chemical attack during my morning subway commute in Tokyo back in March of 1995. It was later revealed to what great lengths Aum had gone to weaponize biological agents such as anthrax and the Ebola virus for other assaults on the Japanese capital, where my wife and I were raising two young daughters. That left an indelible impression on me in ways my high school biology teachers never quite managed.

Years later, in the early 2000s, I became acutely aware of the destructive power of a fast-moving and novel viral respiratory illness called severe acute respiratory syndrome (SARS) that rampaged around Asia and other parts of the world. In the following decade came infectious disease outbreaks of Ebola and Zika that were further confirmation that humanity shares the biosphere with an unseen and pulsating realm of microorganisms. Then came COVID-19, the biggest global health emergency in more than a century.

As the senior executive editor in charge of global business coverage at Bloomberg News, I've had the good fortune to lead some of the best health and science journalists in the business, as we chronicled a global pandemic that killed millions and continues to disrupt lives worldwide. Their award-winning reporting, insights and expertise provided a valuable public service in a worldwide health crisis, and I'm proud to work with every one of them.

I would like to thank Mike Bloomberg, John Micklethwait, Reto Gregori, Heather Harris and John Fraher for their support with this project.

I'm grateful as well to Nick Wallwork, Chris Newson and Chua Hong Koon for seeing the value in an explanatory work aimed at a non-specialist, general reader that explores why microbes are at the center of all of this century's biggest challenges: pandemics, food security, ocean health, biosecurity, climate change, and even space exploration.

Above all, I'd like to express my appreciation to the scientists, technologists, business leaders and government officials who shared their time and expertise in extended interviews, illuminating the finer points of microbiology, biotech, gene editing, infectious diseases, synthetic and computational biology, bioinformatics, biosecurity, among other disciplines.

In particular, I'd like to thank Anthony Fauci, Dennis Carroll, Jonna Mazet, Christopher Mason, Grey Frandsen, Kamran Khan, Nita Madhav, Eric Schmidt, Richard Hatchett, Drew Weissman, Katalin Karikó, David Nabarro, Steffanie Strathdee, Steve Maund, Ian Jepson, Martin Clough, Brian Brazeau, Fotis Fotiadis, David J. Smith, Susan Natali, Will Wieder, Tom Knight, Andrew Weber, James Diggans, Claudius Gros, and Ester Baiget. Their knowledge and perspectives greatly enhanced this book. Any shortcomings are mine alone.

Our relationship with microbes is a paradoxical one, both adversarial and mutually beneficial. If we are going to navigate future microbial threats and realize the vast potential of these amazing life forms, the general public is going to need to be better informed about this invisible realm and more involved in the global response to the next infectious disease outbreak, whenever that day comes.

You don't need an advanced degree in microbiology to appreciate how essential these creatures are to planetary health, or perhaps even be slightly in awe of their staying power over billions of years. Ignoring, or worse politicizing, the risks they pose is no longer an option for humanity. After living through a pandemic that exacted such a staggering loss of human life worldwide, how could it be?

London, March 2022

References

Introduction

1. Danzig, R., Sageman, M., Leighton, T., Hough, L., Yuki, H., Kotani, R. & Hosford, Z. M. (2012, December 1). *Aum Shinrikyo: Insights into how terrorists develop biological and chemical weapons* (2nd ed.). Center for a New American Security. www.jstor.org/stable/resrep06323

2. Fleischmann, R. D., Adams, M. D., White, O., Clayton, R. A., Kirkness, E. F., Kerlavage, A. R., Bult, C. J., Tomb, J.-F., Dougherty, B. A., Merrick, J. M., McKenney, K., Sutton, G., FitzHugh, W., Fields, C., Gocyne, J. D., Scott, J., Shirley, R., Liu, L.-I., Glodek, A., . . . Venter, J. C. (1995). Whole-genome random sequencing and assembly of *Haemophilus influenzae* Rd. *Science, 269*(5223), 496–512. http://www.jstor.org/stable/2887657

3. Venter, J. C. (2007). *A life decoded: My genome: My life*. Viking Books.

4. Gibson, D. G., Glass, J. I., Lartigue, C., Noskov, V. N., Chuang, R. Y., Algire, M. A., Benders, G. A., Montague, M. G., Ma, L., Moodie, M. M., Merryman, C., Vashee, S., Krishnakumar, R., Assad-Garcia, N., Andrews-Pfannkoch, C., Denisova, E. A., Young, L., Qi, Z. Q., Segall-Shapiro, T. H., . . . Venter, J. C. (2010). Creation of a bacterial cell controlled by a chemically synthesized genome. *Science, 329*(5987), 52–56. https://doi.org/10.1126/science.1190719

5. Pilkington, E. (2007, October 6). I am creating artificial life, declares US gene pioneer. *The Guardian*. https://www.theguardian.com/science/2007/oct/06/genetics.climatechange

6. Jinek, M., Chylinski, K., Fonfara, I., Hauer, M., Doudna, J. A. & Charpentier, E. (2012). A programmable dual-RNA-guided DNA endonuclease in adaptive bacterial immunity. *Science, 337*(6096), 816–821. https://doi.org/10.1126/science.1225829
7. Nobel Prize Outreach AB 2021. (n.d.). *The Nobel Prize in Chemistry 2020.* NobelPrize.org. https://www.nobelprize.org/prizes/chemistry/2020/summary/
8. University of Georgia. (1998, August 25). First-ever scientific estimate of total bacteria on Earth shows far greater numbers than ever known before. *Science Daily*. www.sciencedaily.com/releases/1998/08/980825080732.htm
9. Hug, L. A., Baker, B. J., Anantharaman, K, Brown, C. T., Probst, A. J., Castelle, C. J., Butterfield, C. N., Hernsdorf, A. W., Amano, Y., Ise, K., Suzuki, Y., Dudek, N., Relman, D. A., Finstad, K. M., Amundson, R., Thomas, B. C. & Banfield, J. F. (2016). A new view of the tree of life. *Nature Microbiology, 1*(16048). https://doi.org/10.1038/nmicrobiol.2016.48
10. Microbiology by numbers. (2011). *Nature Reviews Microbiology, 9*(628). https://doi.org/10.1038/nrmicro2644
11. Cavicchioli, R., Ripple, W. J., Timmis, K. N., Azam, F., Bakken, L. R., Baylis, M., Behrenfeld, M. J., Boetius, A., Boyd, P. W., Classen, A. T., Crowther, T. W., Danovaro, R., Foreman, C. M., Huisman, J., Hutchins, D. A., Jansson, J. K., Karl, D. M., Koskella, B., Mark Welch, D. B., . . . Webster, N. S. (2019). Scientists' warning to humanity: Microorganisms and climate change. *Nature Reviews Microbiology, 17*, 569–586. https://doi.org/10.1038/s41579-019-0222-5
12. The Independent Panel for Pandemic Preparedness and Response. (n.d.). *COVID-19: Make it the last pandemic.* https://theindependentpanel.org/wp-content/uploads/2021/05/COVID-19-Make-it-the-Last-Pandemic_final.pdf
13. World Health Organization. (2018). *Managing epidemics: Key facts about major deadly diseases.* https://www.who.int/emergencies/diseases/managing-epidemics-interactive.pdf

14. United Nations: Department of Economic and Social Affairs. (2019, December). How certain are the United Nations global population projections? *Population Facts*, 6. https://www.un.org/en/development/desa/population/publications/pdf/popfacts/PopFacts_2019-6.pdf

15. Koplow, D. (2003). *Smallpox: The fight to eradicate a global scourge*. University of California Press.

16. Imai, M., Watanabe, T., Hatta, M., Das, S. C., Ozawa, M., Shinya, K., Zhong, G., Hanson, A., Katsura, H., Watanabe, S., Li, C., Kawakami, E., Yamada, S., Kiso, M., Suzuki, Y., Maher, E. A., Neumann, G. & Kawaoka, Y. (2012). Experimental adaptation of an influenza H5 HA confers respiratory droplet transmission to a reassortant H5 HA/H1N1 virus in ferrets. *Nature, 486*, 420–428. https://doi.org/10.1038/nature10831

17. Noyce, R. S. & Evans, D. H. (2018). Synthetic horsepox viruses and the continuing debate about dual use research. *PLOS Pathogens, 14*(10), e1007025. https://doi.org/10.1371/journal.ppat.1007025

Chapter 1

1. The White House (2008, June 19). *2018 Presidential Medal of Freedom citations*. https://georgewbush-whitehouse.archives.gov/news/releases/2008/06/20080619-3.html

2. McNeil, D. G., Jr. (2021, January 24). Fauci on what working for trump was really like. *The New York Times*. https://www.nytimes.com/2021/01/24/health/fauci-trump-covid.html?action=click&module=Top%20Stories&pgtype=Homepage

3. Scherer, M. & Dawsey, J. (2020, October 19). Trump attacks 'Fauci and all these idiots,' says public is tired of pandemic, public health restrictions as infection rates rise. *The Washington Post*. https://www.washingtonpost.com/politics/trump-fauci-campaign-biden/2020/10/19/30b2fe58-1226-11eb-82af-864652063d61_story.html

4. Beaumont, P. (2020, November 6). Steve Bannon banned by Twitter for calling for Fauci beheading. *The Guardian*. https://www.theguardian.com/us-news/2020/nov/06/steve-bannon-banned-by-twitter-for-calling-for-fauci-beheading
5. Baker, S. (2021, March 31). Trump advisor Peter Navarro went on a wild rant on Fox News, calling Fauci the 'father' of the coronavirus. *Business Insider*. https://www.businessinsider.com/peter-navarro-trump-advisor-calls-fauci-father-of-coronavirus-fox-news-rant-2021-3?r=US&IR=T
6. Andrews, S. M. & Rowland-Jones, S. (2017). Recent advances in understanding HIV evolution. *F1000 Research, 6*(597). https://doi.org/10.12688/f1000research.10876.1
7. UNAIDS. (n.d.). *Global HIV & AIDS statistics — Fact sheet*. https://www.unaids.org/en/resources/fact-sheet
8. Connors, T., Byhoff, M. & Bremner, B. (2020, December 30). *Contagion of the mind: How the world failed in 2020* [Video]. Bloomberg. https://www.bloomberg.com/news/videos/2020-12-30/contagion-of-the-mind-how-the-world-failed-in-2020-video-kjbn62hj?sref=hYP6YGUZ
9. Seitz, B. M., Aktipis, A., Buss, D. M., Alcock, J., Bloom, P., Gelfand, M., Harris, S., Lieberman, D., Horowitz, B. N., Pinker, S., Wilson D. S. & Haselton, M. G. (2020). The pandemic exposes human nature: 10 evolutionary insights. *Proceedings of the National Academy of Sciences of the United States of America, 117*(45). https://www.pnas.org/content/117/45/27767
10. Schaller M. (2011). The behavioural immune system and the psychology of human sociality. *Philosophical Transactions of the Royal Society B: Biological Sciences, 366*(1583), 3418–3426. https://doi.org/10.1098/rstb.2011.0029
11. Littman R. J. (2009). The Plague of Athens: Epidemiology and paleopathology. *Mount Sinai Journal of Medicine, 76*(5), 456–467. https://pubmed.ncbi.nlm.nih.gov/19787658/
12. Page, D. L. (1953). Thucydides' description of the Great Plague at Athens. *The Classical Quarterly, 3*(3/4), 97–119. http://www.jstor.org/stable/637025

13. Zheng, S. (2020, January 23). Wuhan mayor under pressure to resign over response to coronavirus outbreak. *South China Morning Post.* https://www.scmp.com/news/china/politics/article/3047230/wuhan-mayor-under-pressure-resign-over-response-coronavirus

14. The Independent Panel for Pandemic Preparedness and Response. (n.d.). *COVID-19: Make it the last pandemic.* https://theindependentpanel.org/wp-content/uploads/2021/05/COVID-19-Make-it-the-Last-Pandemic_final.pdf

15. Roser, M., Ochmann, S., Behrens, H., Ritchie H. & Dadonaite B. (2014, June). *Eradication of diseases.* Our World in Data. https://ourworldindata.org/eradication-of-diseases

16. Ritzmann, I. (1998). Judenmord als folge des "Schwarzen Todes": Ein medizin-historischer mythos? [The Black Death as a cause of the massacres of Jews: a myth of medical history?]. *Medizin, Gesellschaft und Geschichte: Jahrbuch des Instituts fur Geschichte der Medizin der Robert Bosch Stiftung, 17,* 101–130. https://pubmed.ncbi.nlm.nih.gov/11625662/

17. Gunderman, R. (2020, March 17). Ten myths about the 1918 flu pandemic. *Smithsonian Magazine.* https://www.smithsonianmag.com/history/ten-myths-about-1918-flu-pandemic-180967810/

18. Saric, I. (2021, September 23). Murders up nearly 30% nationwide in 2020. *Axios.* https://www.axios.com/murder-rate-rose-fbi-30-percent-2020-828855f2-8985-431d-8417-f91984ee55e4.html

19. Brüne, M. & Wilson, D. R. (2020). Evolutionary perspectives on human behavior during the coronavirus pandemic: Insights from game theory. *Evolution, Medicine, and Public Health, 2020*(1), 181–186. https://doi.org/10.1093/emph/eoaa034

20. Micklethwait, J. & Wooldridge, A. (2020). *The wake up call: Why the pandemic has exposed the weakness of the west, and how to fix it.* HarperVia.

21. Gelfand, M. J., Jackson, J. C., Pan, X., Nau, D., Pieper, D., Denison, E., Dagher, M., Van Lange, P. A. M., Chiu, C.-Y., & Wang, M. (2021). The relationship between cultural tightness–looseness and COVID-19 cases and deaths: A global analysis. *The Lancet Planetary Health. 5*(3),

E135–E144. https://www.thelancet.com/journals/lanplh/article/PIIS2542-5196(20)30301-6/fulltext

22. Jamieson, K. H. & Albarracín, D. (2020). The relation between media consumption and misinformation at the outset of the SARS-CoV-2 pandemic in the U.S. *Harvard Kennedy School (HKS Misinformation Review, 1.* https://misinforeview.hks.harvard.edu/article/the-relation-between-media-consumption-and-misinformation-at-the-outset-of-the-sars-cov-2-pandemic-in-the-us/

Chapter 2

1. Sender, R., Fuchs, S. & Milo, R. (2016). Are we really vastly outnumbered? Revisiting the ratio of bacterial to host cells in humans. *Cell, 164*(3), 337–340. https://doi.org/10.1016/j.cell.2016.01.013
2. Hood L. (2012). Tackling the microbiome. *Science, 336*(6086), 1209. https://doi.org/10.1126/science.1225475
3. United Nations, Department of Economic and Social Affairs, Population Division. (2018, December 1). *The world's cities in 2018 data booklet.* https://www.un.org/development/desa/pd/content/worlds-cities-2018-data-booklet
4. Afshinnekoo, E., Meydan, C., Chowdhury, S., Jaroudi, D., Boyer, C., Bernstein, N., Maritz, J. M., Reeves, D., Gandara, J., Chhangawala, S., Ahsanuddin, S., Simmons, A., Nessel, T., Sundaresh, B., Pereira, E., Jorgensen, E., Kolokotronis, S.-O., Kirchberger, N., Garcia, I., . . . Mason, C. E. (2015). Geospatial resolution of human and bacterial diversity with city-scale metagenomics. *Cell Systems, 1*(1), 72–87. https://doi.org/10.1016/j.cels.2015.01.001
5. Danko, D., Bezdan, D., Afshin, E. E., Ahsanuddin, S., Bhattacharya, C., Butler, D. J., Chng, K. R., Donnellan, D., Hecht, J., Jackson, K., Kuchin, K., Karasikov, M., Lyons, A., Mak, L., Meleshko, D., Mustafa, H., Mutai, B., Neches, R. Y., Ng, A., . . . The International MetaSUB Consortium (2021). A global metagenomic map of urban microbiomes

and antimicrobial resistance. *Cell, 184*(13), 3376–3393. https://doi.org/10.1016/j.cell.2021.05.002

6. Wellcome Trust & UK Government. (2016, May). *Tackling drug-resistant infections globally: Final report and recommendations. The review on antimicrobial resistance chaired by Jim O'Neill.* https://wellcomecollection.org/works/thvwsuba/items

7. Garrett-Bakelman, F. E., Darshi, M., Green, S. J., Gur, R. C., Lin, L., Macias, B. R., McKenna, M. J., Meydan, C., Mishra, T., Nasrini, J., Piening, B. D., Rizzardi, L. F., Sharma, K., Siamwala, J. H., Taylor, L., Vitaterna, M. H., Afkarian, M., Afshinnekoo, E., Ahadi, S., . . . Turek, F. W. (2019). The NASA Twins Study: A multidimensional analysis of a year-long human spaceflight. *Science, 364*(6436). https://doi.org/10.1126/science.aau8650

8. Mason, C. E. (2021). *The next 500 years: Engineering life to reach new worlds.* MIT Press.

9. Gopinath, G., Brooks, P. K., Nabar, M. & Anspach, R. (2021, January 28). *Transcript of the world economic outlook update press briefing.* International Monetary Fund. https://www.imf.org/en/News/Articles/2021/01/28/tr012621-transcript-of-the-world-economic-outlook-update-press-briefing

10. Carroll, D., Daszak, P., Wolfe, N. D., Gao, G. F., Morel, C. M., Morzaria, S., Pablos-Méndez, A., Tomori, O. & Mazet, J. A. K. (2018). The Global Virome Project. *Science, 359*(6378), 872–874. https://doi.org/10.1126/science.aap7463

11. PREDICT Consortium. (2020). *Advancing global health security at the frontiers of disease emergence.* One Health Institute, University of California, Davis. https://ohi.vetmed.ucdavis.edu/sites/g/files/dgvnsk5251/files/inline-files/PREDICT%20LEGACY%20-%20FINAL%20FOR%20WEB%20-compressed_0.pdf

12. World Health Organization. (2012, March 11). *Influenza: H5N1.* https://www.who.int/news-room/q-a-detail/influenza-h5n1

13. World Health Organization. (2018, May 30). *Nipah virus.* https://www.who.int/news-room/fact-sheets/detail/nipah-virus

14. Grange, Z. L., Goldstein, T., Johnson, C. K., Anthony, S., Gilardi, K., Daszak, P., Olival, K. J., O'Rourke, T., Murray, S., Olson, S. H., Togami, E., Vidal, G., Expert Panel, PREDICT Consortium, Mazet, J. A. K. & University of Edinburgh Epigroup members those who wish to remain anonymous. (2021). Ranking the risk of animal-to-human spillover for newly discovered viruses. *Proceedings of the National Academy of Sciences of the United States of America, 118*(15), e2002324118. https://doi.org/10.1073/pnas.2002324118

15. Kraemer, M. U. G., Reiner, R. C., Jr., Brady, O. J., Messina, J. P., Gilbert, M., Pigott, D. M., Yi, D., Johnson, K., Earl, L., Marczak, L. B., Shirude, S., Weaver, N. D., Bisanzio, D., Perkins, T. A., Lai, S., Lu, X., Jones, P., Coelho, G. E., Carvalho, R. G., . . . Golding, N. (2019). Past and future spread of the arbovirus vectors *Aedes aegypti* and *Aedes albopictus*. *Nature Microbiology, 4*, 854–863. https://doi.org/10.1038/s41564-019-0376-y

Chapter 3

1. Low, D.E. (2004). SARS: Lessons from Toronto. In: S. Knobler, A. Mahmoud, S. Lemon, A. Mack, L. Sivitz & K. Oberholtzer (Eds.). *Learning from SARS: Preparing for the next disease outbreak: Workshop summary.* The National Academies Press. https://www.ncbi.nlm.nih.gov/books/NBK92467/

2. Government of Canada. (2021, October 29). *COVID-19 daily epidemiology update*. https://health-infobase.canada.ca/covid-19/epidemiological-summary-covid-19-cases.html

3. Lazer, D., Kennedy, R., King, G., & Vespignani, A. (2014). The parable of Google Flu: Traps in big data analysis. *Science, 343*(6176), 1203–1205. https://doi.org/10.1126/science.1248506

4. Boehme, C., Hannay, E. & Pai, M. (2021). Promoting diagnostics as a global good. *Nature Medicine, 27*, 367–368. https://doi.org/10.1038/s41591-020-01215-3

5. McCarthy, J., Minsky, M. L., Rochester, N. & Shannon, C. E. (1956). *A proposal for the Dartmouth summer research project on artificial intelligence.* Dartmouth University. https://250.dartmouth.edu/highlights/artificial-intelligence-ai-coined-dartmouth
6. Bremner, B., Langreth, R., & Paton, J. (2020, February 6) Man vs. microbe: We're not ready for the next global virus outbreak. *Bloomberg Businessweek*. https://www.bloomberg.com/news/articles/2020-02-06/forget-coronavirus-world-isn-t-ready-for-next-global-outbreak?sref=hY-P6YGUZ
7. ZOE COVID Study. (n.d.). https://covid.joinzoe.com
8. Mozzie Monitors. (n.d.). http://mozziemonitors.com/index.php
9. Newcastle University. (n.d.). *Detecting disease via mobile phone*. https://www.ncl.ac.uk/research/impact/casestudies/detecting-mobile/#discover-more
10. Bateman, T. (2020, May 12). Coronavirus: Israel turns surveillance tools on itself. *BBC News*. https://www.bbc.co.uk/news/world-middle-east-52579475
11. Fan, V. Y., Jamison, D. T. & Summers, L. H. (2018). Pandemic risk: How large are the expected losses? *Bulletin of the World Health Organization, 96*(2), 129–134. https://doi.org/10.2471/BLT.17.199588
12. Beck, M., Asenova, D. & Dickson, G. (2005). Public administration, science, and risk assessment: A case study of the U.K. bovine spongiform encephalopathy crisis. *Public Administration Review, 65*(4), 396–408. http://www.jstor.org/stable/3542637
13. Wolfe, N. (2012). *The viral storm: The dawn of a new pandemic age.* St. Martin's Griffin.
14. Rossi, B. (2019, April 10). *The metabiota story*. Medium. https://medium.com/@billrossi/the-metabiota-story-d553f41d03bd
15. Dyer, O. (2021). Covid-19: Study claims real global deaths are twice official figures. *The BMJ, 373*, n1188. https://doi.org/10.1136/bmj.n1188

Chapter 4

1. Silver, D., Huang, A., Maddison, C. J., Guez, A., Sifre, L., van den Driessche, G., Schrittwieser, J., Antonoglou, I., Panneershelvam, V., Lanctot, M., Dieleman, S., Grewe, D., Nham, J., Kalchbrenner, N., Sutskever, I., Lillicrap, T., Leach, M., Kavukcuoglu, K., Graepel, T. & Hassabis, D. (2016). Mastering the game of Go with deep neural networks and tree search. *Nature, 529*, 484–489. https://doi.org/10.1038/nature16961
2. Tunyasuvunakool, K., Adler, J., Wu, Z., Green, T., Zielinski, M., Žídek, A., Bridgland, A., Cowie, A., Meyer, C., Laydon, A., Velankar, S., Kleywegt, G. J., Bateman, A., Evans, R., Pritzel, A., Figurnov, M., Ronneberger, O., Bates, R., Kohl, S. A. A., . . . Hassabis, D. (2021). Highly accurate protein structure prediction for the human proteome. *Nature*, *596*, 590–596. https://doi.org/10.1038/s41586-021-03828-1
3. Jumper, J., Tunyasuvunakool, K., P Kohli, P., Hassabis, D. & the AlphaFold Team. (2020, August 4). *Computational predictions of protein structures associated with COVID-19* (Version 3). DeepMind. https://deepmind.com/research/open-source/computational-predictions-of-protein-structures-associated-with-COVID-19
4. Burton, T. D. & Eyre, N. S. (2021). Applications of deep mutational scanning in virology. *Viruses*, *13*(6), 1020. https://doi.org/10.3390/v13061020
5. Higgins, M. K. (2021). Can we AlphaFold our way out of the next pandemic? *Journal of Molecular Biology*, *433*(20). https://doi.org/10.1016/j.jmb.2021.167093
6. Lindmeier, C. (2020, May 8). *Commemorating smallpox eradication — A legacy of hope, for COVID-19 and other diseases*. World Health Organization. https://www.who.int/news/item/08-05-2020-commemorating-smallpox-eradication-a-legacy-of-hope-for-covid-19-and-other-diseases
7. Strouhal, E. (1996). Traces of a smallpox epidemic in the family of Ramesses V of the Egyptian 20th dynasty. *Anthropologie (1962-)*, *34*(3), 315–319. http://www.jstor.org/stable/44601512

8. Riedel S. (2005). Edward Jenner and the history of smallpox and vaccination. *Baylor University Medical Center Proceedings*, *18*(1), 21–25. https://doi.org/10.1080/08998280.2005.11928028
9. Juskewitch, J. E., Tapia, C. J. & Windebank, A. J. (2010). Lessons from the Salk polio vaccine: Methods for and risks of rapid translation. *Clinical and Translational Science*, *3*(4), 182–185. https://www.ncbi.nlm.nih.gov/pmc/articles/PMC2928990/
10. Jacobs, J. & Armstrong, D. (2020, April 29). Trump's 'Operation Warp Speed' aims to rush coronavirus vaccine. *Bloomberg News*. https://www.bloomberg.com/news/articles/2020-04-29/trump-s-operation-warp-speed-aims-to-rush-coronavirus-vaccine?sref=hYP6YGUZ
11. Jacob, F. & Monod, J. (1961). Genetic regulatory mechanisms in the synthesis of proteins. *Journal of Molecular Biology*, *3*(3), 318–356. https://doi.org/10.1016/s0022-2836(61)80072-7
12. Karikó, K., Buckstein, M., Ni, H. & Weissman, D. (2005). Suppression of RNA recognition by toll-like receptors: The impact of nucleoside modification and the evolutionary origin of RNA. *Immunity*, *23*(2), 165–175. https://doi.org/10.1016/j.immuni.2005.06.008
13. Bloomberg QuickTake. (2021, April 9). *How covid vaccine tech could fight cancer soon* [Video]. YouTube. https://www.youtube.com/watch?v=b3hWEC553sU
14. Baker, S., Lauerman, J. & Paton, J. (2020, March 30). How top scientists are racing to beat the coronavirus. *Bloomberg News*. https://www.bloomberg.com/news/features/2020-03-30/coronavirus-the-scientists-hunting-for-treatments-vaccine?sref=hYP6YGUZ
15. Wighton, K. (2018, December 10). *Tailor-made flu and Disease X vaccines to be created in $8 million project*. Imperial College London. https://www.imperial.ac.uk/news/189447/tailor-made-disease-vaccines-created-million-project/
16. Watterson, D., Wijesundara, D. K., Modhiran, N., Mordant, F. L., Li, Z., Avumegah, M. S., McMillan, C. L. D., Lackenby, J., Guilfoyle, K., van

Amerongen, G., Stittelaar, K., Cheung, S. T. M., Bibby, S., Daleris, M., Hoger, K., Gillard, M., Radunz, E., Jones, M. L., Hughes, K., . . . Chappell, K. J. (2021). Preclinical development of a molecular clamp-stabilised subunit vaccine for severe acute respiratory syndrome coronavirus 2. *Clinical & Translational Immunology*, *10*(4), e1269. https://doi.org/10.1002/cti2.1269

17. Nazar, S. & Pieters, T. (2021). *Plandemic* revisited: A product of planned disinformation amplifying the COVID-19 "infodemic". *Frontiers in Public Health*, *9*, 649930. https://doi.org/10.3389/fpubh.2021.649930

18. Zola Matuvanga, T., Johnson, G., Larivière, Y., Esanga Longomo, E., Matangila, J., Maketa, V., Lapika, B., Mitashi, P., Mckenna, P., De Bie, J., Van Geertruyden, J., Van Damme, P. & Muhindo Mavoko, H. (2021). Use of iris scanning for biometric recognition of healthy adults participating in an Ebola vaccine trial in the Democratic Republic of the Congo: Mixed methods study. *Journal of Medical Internet Research*, *23*(8), e28573. https://www.jmir.org/2021/8/e28573

Chapter 5

1. The White House. (2021, September 3). *Fact sheet: Biden Administration to transform capabilities for pandemic preparedness*. https://www.whitehouse.gov/briefing-room/statements-releases/2021/09/03/fact-sheet-biden-administration-to-transform-capabilities-for-pandemic-preparedness/

2. Lindmeier, C., Kautz, H. & Rohrbach, L. (2021, May 5). *WHO, Germany launch new global hub for pandemic and epidemic intelligence*. World Health Organization. https://www.who.int/news/item/05-05-2021-who-germany-launch-new-global-hub-for-pandemic-and-epidemic-intelligence

3. Finch, C. E. (2010). Evolution of the human lifespan and diseases of aging: Roles of infection, inflammation, and nutrition. *Proceedings of the National Academy of Sciences of the United States of America*, *107*(Suppl. 1), 1718–1724. https://doi.org/10.1073/pnas.0909606106

4. Huber, V. (2006). The unification of the globe by disease? The International Sanitary Conferences on Cholera, 1851–1894. *The Historical Journal, 49*(2), 453–476. doi:10.1017/S0018246X06005280

5. Tulchinsky, T. H. (2018). *Case studies in public health.* Academic Press. https://doi.org/10.1016/B978-0-12-804571-8.00017-2

6. Lippi, D. & Gotuzzo, E. (2014). The greatest steps towards the discovery of *Vibrio cholerae. Clinical Microbiology and Infection, 20*(3),191–195. https://doi.org/10.1111/1469-0691.12390

7. Blevins, S. M. & Bronze, M. S. (2010). Robert Koch and the 'golden age' of bacteriology. *International Journal of Infectious Diseases, 14*(9), E744–E751. https://doi.org/10.1016/j.ijid.2009.12.003

8. Roser, M., Ortiz-Ospina, E. & Ritchie, H. (2013). *Life expectancy.* Our World in Data. https://ourworldindata.org/life-expectancy

9. World Health Organization. (1946, July 22). *Constitution of the World Health Organization.* https://treaties.un.org/doc/Treaties/1948/04/19480407%2010-51%20PM/Ch_IX_01p.pdf

10. Belongia, E. A. & Naleway, A. L. (2003). Smallpox vaccine: The good, the bad, and the ugly. *Clinical Medicine & Research, 1*(2), 87–92. https://doi.org/10.3121/cmr.1.2.87

11. Ruger, J. P., & Yach, D. (2009). The global role of the World Health Organization. *Global Health Governance, 2*(2), 1–11. https://www.ncbi.nlm.nih.gov/pmc/articles/PMC3981564/

12. Moon, S., Sridhar, D., Pate, M. A., Jha, A. K., Clinton, C., Delaunay, S., Edwin, V., Fallah, M., Fidler, D. P., Garrett, L., Goosby, E., Gostin, L. O., Heymann, D. L., Lee, K., Leung, G. M., Morrison, J. S., Saavedra, J., Tanner, M., Leigh, J. A., . . . Piot, P. (2015). Will Ebola change the game? Ten essential reforms before the next pandemic. The report of the Harvard-LSHTM Independent Panel on the Global Response to Ebola. *The Lancet, 386*(10009), 2204–2221. https://doi.org/10.1016/S0140-6736(15)00946-0

13. Sirleaf, E. J. & Clark, H. (2021). Report of the Independent Panel for Pandemic Preparedness and Response: Making COVID-19 the last pandemic. *The Lancet, 398*(10295), 101–103. https://doi.org/10.1016/S0140-6736(21)01095-3

14. World Health Organization (WHO). (2016, November 8). *WHO: Dr David Nabarro at the director-general candidates forum* [Video]. YouTube. https://www.youtube.com/watch?v=aLj9vf1ilfw

15. Schemm, P. (2020, November 19). Ethiopia's military chief calls WHO head Tedros a criminal supporting a rebel region. *The Washington Post.* https://www.washingtonpost.com/world/2020/11/19/ethiopia-who-tedros-criminal-military-tigray/

16. World Health Organization [@who]. (2020, January 14). *Preliminary investigations conducted by the Chinese authorities have found no clear evidence of human-to-human transmission of the novel #coronavirus (2019-nCoV) identified in #Wuhan, #China*. [Tweet; attached image]. Twitter. https://twitter.com/who/status/1217043229427761152?lang=en

17. Lauerman, J. (2020, February 24). As pandemic looms, world's top disease fighter engages Xi: Tedros Adhanom Ghebreyesus is racing against time to stop the new coronavirus from becoming a global crisis. *Bloomberg News.* https://www.bloomberg.com/news/features/2020-02-23/coronavirus-news-who-director-general-races-against-time?sref=hYP6YGUZ

18. United Nations. (2003, August 22). Eyewitness gives account of attack on UN headquarters in Baghdad. *UN News*. https://news.un.org/en/story/2003/08/77452-eyewitness-gives-account-attack-un-headquarters-baghdad

19. The BMJ Opinion. (2017, May 19). *David Nabarro: The WHO must change, and I am the right person to deliver that change*. https://blogs.bmj.com/bmj/2017/05/19/david-nabarro-the-who-must-change-and-i-am-the-right-person-to-deliver-that-change/

20. McKinsey & Company (2021, June 10). *Second time right*. https://www.mckinsey.com/featured-insights/coronavirus-leading-through-the-crisis/charting-the-path-to-the-next-normal/second-time-right

Chapter 6

1. Sahel, J.-A., Boulanger-Scemama, E., Pagot, C., Arleo, A., Galluppi, F., Martel, J. N., Esposti, S. D., Delaux, A., de Saint Aubert, J.-B., de Montleau, C., Gutman, E., Audo, I., Duebel, J., Picaud, S., Dalkara, D., Blouin, L., Taiel, M. & Roska, B. (2021). Partial recovery of visual function in a blind patient after optogenetic therapy. *Nature Medicine, 27*, 1223–1229. https://doi.org/10.1038/s41591-021-01351-4

2. Boyden, E. S. (2011). A history of optogenetics: the development of tools for controlling brain circuits with light. *F1000 Biology Reports, 3*(11). https://doi.org/10.3410/B3-11

3. GenSight Biologics. (2021, June 4). *KOL webinar on the Nature Medicine Case Report: Visual recovery after GS030 optogenetic treatment*. https://www.gensight-biologics.com/2021/05/28/kol-webinar-on-the-nature-medicine-case-report-visual-recovery-after-gs030-optogenetic-treatment/

4. MacDougall, R. (2012, June 13). *NIH Human Microbiome Project defines normal bacterial makeup of the body*. National Institutes of Health. https://www.nih.gov/news-events/news-releases/nih-human-microbiome-project-defines-normal-bacterial-makeup-body

5. Tierney, B. T., Yang, Z., Luber, J. M., Beaudin, M., Wibowo, M. C., Baek, C., Mehlenbacher, E., Patel, C. J. & Kostic, A. D. (2019). The landscape of genetic content in the gut and oral human microbiome. *Cell Host & Microbe, 26*(2), 283–295. https://doi.org/10.1016/j.chom.2019.07.008

6. Sender, R., Fuchs, S., & Milo, R. (2016). Revised estimates for the number of human and bacteria cells in the body. *PLOS Biology, 14*(8), e1002533. https://doi.org/10.1371/journal.pbio.1002533

7. Shao, Y., Forster, S. C., Tsaliki, E., Vervier, K., Strang, A., Simpson, N., Kumar, N., Stares, M. D., Rodger, A., Brocklehurst, P., Field, N. & Lawley, T. D. (2019). Stunted microbiota and opportunistic pathogen colonization in caesarean-section birth. *Nature, 574*, 117–121. https://doi.org/10.1038/s41586-019-1560-1

8. Zuo, T., Kamm, M. A., Colombel, J.-F. & Ng, S. C. (2018). Urbanization and the gut microbiota in health and inflammatory bowel disease. *Nature Reviews Gastroenterology & Hepatology, 15*, 440–452. https://doi.org/10.1038/s41575-018-0003-z

9. Tun, H. M., Konya, T., Takaro, T. K., Brook, J. R., Chari, R., Field, C. J., Guttman, D. S., Becker, A. B., Mandhane, P. J., Turvey, S. E., Subbarao, P., Sears, M. R., Scott, J. A., Kozyrskyj, A. L. & the CHILD Study Investigators (2017). Exposure to household furry pets influences the gut microbiota of infants at 3–4 months following various birth scenarios. *Microbiome, 5*(40). https://doi.org/10.1186/s40168-017-0254-x

10. de la Cuesta-Zuluaga, J., Kelley, S. T., Chen, Y., Escobar, J. S., Mueller, N. T., Ley, R. E., McDonald, D., Huang, S., Swafford, A. D., Knight, R. & Thackray, V. G. (2019). Age- and sex-dependent patterns of gut microbial diversity in human adults. *mSystems, 4*(4), e00261-19. https://doi.org/10.1128/mSystems.00261-19

11. Ursell, L. K., Metcalf, J. L., Parfrey, L. W. & Knight, R. (2012). Defining the human microbiome. *Nutrition Reviews, 70*(Suppl. 1), S38–S44. https://doi.org/10.1111/j.1753-4887.2012.00493.x

12. Vlachos, C., Gaitanis, G., Katsanos, K. H., Christodoulou, D. K., Tsianos, E. & Bassukas, I. D. (2016). Psoriasis and inflammatory bowel disease: Links and risks. *Psoriasis: Targets and Therapy, 2016*(6), 73–92. https://doi.org/10.2147/PTT.S85194

13. Luczynski, P., McVey Neufeld, K. A., Oriach, C. S., Clarke, G., Dinan, T. G. & Cryan, J. F. (2016). Growing up in a bubble: Using germ-free animals to assess the influence of the gut microbiota on brain and behavior. *The International Journal of Neuropsychopharmacology, 19*(8), 1–17. https://doi.org/10.1093/ijnp/pyw020

14. Fülling, C., Dinan, T. G. & Cryan, J. F. (2019). Gut microbe to brain signaling: What happens in vagus. . . *Neuron, 101*(6), 998–1002. https://doi.org/10.1016/j.neuron.2019.02.008

15. Chudzik, A., Orzyłowska, A., Rola, R. & Stanisz, G. J. (2021). Probiotics, prebiotics and postbiotics on mitigation of depression symptoms:

Modulation of the brain-gut-microbiome axis. *Biomolecules, 11*(7), 1000. https://doi.org/10.3390/biom11071000

16. Alcock, J., Maley, C. C. & Aktipis, C. A. (2014). Is eating behavior manipulated by the gastrointestinal microbiota? Evolutionary pressures and potential mechanisms. *BioEssays, 36*(10), 940–949. https://doi.org/10.1002/bies.201400071

17. Zarocostas, J. (2010). Global cancer cases and deaths are set to rise by 70% in next 20 years. *The BMJ, 340*, c3041. https://doi.org/10.1136/bmj.c3041

18. McCarthy, E. F. (2006). The toxins of William B. Coley and the treatment of bone and soft-tissue sarcomas. *Iowa Orthopaedic Journal, 26*, 154–158. https://pubmed.ncbi.nlm.nih.gov/16789469/

19. Wroblewski, L. E., Peek, R. M., Jr. & Wilson, K. T. (2010). *Helicobacter pylori* and gastric cancer: Factors that modulate disease risk. *Clinical Microbiology Reviews, 23*(4), 713–739. https://doi.org/10.1128/CMR.00011-10

20. Poore, G. D., Kopylova, E., Zhu, Q., Carpenter, C., Fraraccio, S., Wandro, S., Kosciolek, T., Janssen, S., Metcalf, J., Song, S. J., Kanbar, J., Miller-Montgomery, S., Heaton, R., Mckay, R., Patel, S. P., Swafford, A. D. & Knight, R. (2020). Microbiome analyses of blood and tissues suggest cancer diagnostic approach. *Nature, 579*(7800), 567–574. https://doi.org/10.1038/s41586-020-2095-1

21. Duong, M. T.-Q., Qin, Y., You, S.-H. & Min, J.-J. (2019). Bacteria-cancer interactions: Bacteria-based cancer therapy. *Experimental & Molecular Medicine, 51*, 1–15. https://doi.org/10.1038/s12276-019-0297-0

22. Sedighi, M., Zahedi Bialvaei, A., Hamblin, M. R., Ohadi, E., Asadi, A., Halajzadeh, M., Lohrasbi, V., Mohammadzadeh, N., Amiriani, T., Krutova, M., Amini, A. & Kouhsari, E. (2019). Therapeutic bacteria to combat cancer; current advances, challenges, and opportunities. *Cancer Medicine, 8*(6), 3167–3181. https://doi.org/10.1002/cam4.2148

23. Antonelli, A. C., Binyamin, A., Hohl, T. M., Glickman, M. S., & Redelman-Sidi, G. (2020). Bacterial immunotherapy for cancer induces CD4-dependent tumor-specific immunity through tumor-intrinsic

interferon-γ signaling. *Proceedings of the National Academy of Sciences of the United States of America, 117*(31), 18627–18637. https://doi.org/10.1073/pnas.2004421117

24. Lauté-Caly, D. L., Raftis, E. J., Cowie, P., Hennessy, E., Holt, A., Panzica, D. A., Sparre, C., Minter, B., Stroobach, E. & Mulder I. E. (2019). The flagellin of candidate live biotherapeutic *Enterococcus gallinarum* MRx0518 is a potent immunostimulant. *Scientific Reports, 9*(801). https://doi.org/10.1038/s41598-018-36926-8
25. Tanoue, T., Morita, S., Plichta, D. R., Skelly, A. N., Suda, W., Sugiura, Y., Narushima, S., Vlamakis, H., Motoo, I., Sugita, K., Shiota, A., Takeshita, K., Yasuma-Mitobe, K., Riethmacher, D., Kaisho, T., Norman, J. M., Mucida, D., Suematsu, M., Yaguchi, T., . . . Honda, K. (2019). A defined commensal consortium elicits CD8 T cells and anti-cancer immunity. *Nature, 565*(7741), 600–605. https://doi.org/10.1038/s41586-019-0878-z

Chapter 7

1. Microbiology by numbers. (2011). *Nature Reviews Microbiology, 9*(628). https://doi.org/10.1038/nrmicro2644
2. Van Valen, L. (1973). A new evolutionary law. *Evolutionary Theory 1*, 1–30. https://www.mn.uio.no/cees/english/services/van-valen/evolutionary-theory/volume-1/vol-1-no-1-pages-1-30-l-van-valen-a-new-evolutionary-law.pdf
3. Jinek, M., Chylinski, K., Fonfara, I., Hauer, M., Doudna, J. A. & Charpentier, E. (2012). A programmable dual-RNA-guided DNA endonuclease in adaptive bacterial immunity. *Science, 337*(6096), 816–821. https://doi.org/10.1126/science.1225829
4. Nobel Prize Outreach AB 2021. (n.d.). *Advanced information*. NobelPrize.org. https://www.nobelprize.org/prizes/chemistry/2020/advanced-information/
5. Regalado, A. (2018, November 25). Chinese scientists are creating CRISPR babies. *MIT Technology Review*. https://www.technologyreview.

com/2018/11/25/138962/exclusive-chinese-scientists-are-creating-crispr-babies/

6. Sriram, A., Kalanxhi, E., Kapoor, G., Craig, J., Balasubramanian, R., Brar, S., Criscuolo, N., Hamilton, A., Klein, E., Tseng, K., Van Boeckel, T. & Laxminarayan, R. (2021). *The state of the world's antibiotics 2021: A global analysis of antimicrobial resistance and its drivers*. Center for Disease Dynamics, Economics & Policy, Washington, D.C. https://cddep.org/wp-content/uploads/2021/02/The-State-of-the-Worlds-Antibiotics-in-2021.pdf
7. Chaib, F., Butler, J., Hwang, S. & World Organisation for Animal Health (OIE). (2019, April 29). *New report calls for urgent action to avert antimicrobial resistance crisis: International organizations unite on critical recommendations to combat drug-resistant infections and prevent staggering number of deaths each year*. World Health Organization. https://www.who.int/news/item/29-04-2019-new-report-calls-for-urgent-action-to-avert-antimicrobial-resistance-crisis
8. Hendrix, R. W., Smith, M. C., Burns, R. N., Ford, M. E. & Hatfull, G. F. (1999). Evolutionary relationships among diverse bacteriophages and prophages: All the world's a phage. *Proceedings of the National Academy of Sciences of the United States of America*, *96*(5), 2192–2197. https://doi.org/10.1073/pnas.96.5.2192
9. Cohen, K. & Hatfull, G. (Speakers) [NTMir]. (2021, February 10). *Phage (bacteriophage) therapy* [Video]. YouTube. https://www.youtube.com/watch?v=IPxz4Jq-i6Q
10. Institut Pasteur. (2007). On an invisible microbe antagonistic toward dysenteric bacilli: Brief note by Mr. F. D'Herelle, presented by Mr. Roux. *Research in Microbiology*, *158*(7), 553–554. https://doi.org/10.1016/j.resmic.2007.07.005
11. Wellcome Trust. (2020, January 21). *Why is it so hard to develop new antibiotics?* https://wellcome.org/news/why-is-it-so-hard-develop-new-antibiotics
12. Wellcome Trust & UK Government. (2014, December). *Antimicrobial resistance: Tackling a crisis for the health and wealth of nations. The review on antimicrobial resistance chaired by Jim O'Neill*. https://amr-review.org/

sites/default/files/AMR%20Review%20Paper%20-%20Tackling%20a%20crisis%20for%20the%20health%20and%20wealth%20of%20nations_1.pdf

13. El Haddad, L., Harb, C. P., Gebara, M. A., Stibich, M. A. & Chemaly, R. F. (2019). A systematic and critical review of bacteriophage therapy against multidrug-resistant ESKAPE organisms in humans. *Clinical Infectious Diseases, 69*(1), 167–178. https://doi.org/10.1093/cid/ciy947

14. Dąbrowska, K. (2019). Phage therapy: What factors shape phage pharmacokinetics and bioavailability? Systematic and critical review. *Medicinal Research Reviews, 39*(5), 2000–2025. https://doi.org/10.1002/med.21572

Chapter 8

1. Turner, B. L., II. & Sabloff, J. A. (2012). Classic Period collapse of the Central Maya Lowlands: Insights about human-environment relationships for sustainability. *Proceedings of the National Academy of Sciences of the United States of America, 109*(35), 13908–13914. https://www.pnas.org/content/109/35/13908

2. United Nations, Department of Economic and Social Affairs, Population Division. (2019). *World population prospects 2019.* https://population.un.org/wpp/Publications/Files/WPP2019_Highlights.pdf

3. Food and Agricultural Organization of the United Nations. (2011). *The state of the world's land and water resources for food and agriculture: Managing systems at risk.* Earthscan. http://www.fao.org/3/i1688e/i1688e.pdf

4. Food and Agricultural Organization of the United Nations. (2020, May 7). *Land use in agriculture by the numbers.* http://www.fao.org/sustainability/news/detail/en/c/1274219/

5. Peters, G. & Woolley, J. T. (1934, September 6). *Franklin D. Roosevelt: Fireside chat.* The American Presidency Project. https://www.presidency.ucsb.edu/documents/fireside-chat-18

6. United Nations. (2019, June 12). *António Guterres (UN Secretary-General) on World Day to combat desertification* [Video]. YouTube. https://youtu.be/mmZONFMMXqU

7. Thaler, E. A., Larsen, I. J. & Yu, Q. (2021). The extent of soil loss across the US Corn Belt. *Proceedings of the National Academy of Sciences of the United States of America, 118*(8), e1922375118. https://doi.org/10.1073/pnas.1922375118

8. Global Environment Facility. (n.d.). *Land degradation*. https://www.thegef.org/topics/land-degradation

9. Montanarella, L., Scholes, R. & Brainich, A. (Eds.). (2018, March 24). *The IPBES assessment report on land degradation and restoration*. Intergovernmental Science-Policy Platform on Biodiversity and Ecosystem Services. https://doi.org/10.5281/zenodo.3237392

10. The Business Research Company. (2021). *Organic food global market report 2021*. https://www.thebusinessresearchcompany.com/report/organic-food-global-market-report

11. Diab, A. & Adams, G. M. (2021, February 23). *ESG assets may hit $53 trillion by 2025, a third of global AUM*. Bloomberg Intelligence. https://www.bloomberg.com/professional/blog/esg-assets-may-hit-53-trillion-by-2025-a-third-of-global-aum/

12. IPCC. (2019). *Climate change and land: An IPCC special report on climate change, desertification, land degradation, sustainable land management, food security, and greenhouse gas fluxes in terrestrial ecosystems* [P.R. Shukla, J. Skea, E. Calvo Buendia, V. Masson-Delmotte, H.-O. Pörtner, D. C. Roberts, P. Zhai, R. Slade, S. Connors, R. van Diemen, M. Ferrat, E. Haughey, S. Luz, S. Neogi, M. Pathak, J. Petzold, J. Portugal Pereira, P. Vyas, E. Huntley, K. Kissick, M. Belkacemi, J. Malley (Eds.)]. https://www.ipcc.ch/site/assets/uploads/sites/4/2021/07/210714-IPCCJ7230-SRCCL-Complete-BOOK-HRES.pdf

13. General Mills. (2020, September 21). *General Mills to reduce absolute greenhouse gas emissions by 30% across its full value chain over next decade.*

https://investors.generalmills.com/press-releases/press-release-details/2020/General-Mills-to-Reduce-Absolute-Greenhouse-Gas-Emissions-by-30-Across-its-Full-Value-Chain-Over-Next-Decade/default.aspx

14. Gillespie, T., Warren, H. & Randall, T. (2020, December 15). Time's up on corporate America's 2020 climate goals. Here's the results. *Bloomberg Green.* https://www.bloomberg.com/graphics/2020-company-emissions-pledges/?sref=hYP6YGUZ

15. Pobiner, B. (2013). Evidence for meat-eating by early humans. *Nature Education*, *4*(6), 1. https://www.nature.com/scitable/knowledge/library/evidence-for-meat-eating-by-early-humans-103874273/

16. Thornton, A. (2019, February 8). *This is how many animals we eat each year.* World Economic Forum. https://www.weforum.org/agenda/2019/02/chart-of-the-day-this-is-how-many-animals-we-eat-each-year/

17. Wolf, J. (2019). *Microbes, heme, and Impossible Burgers with Pat Brown* [Audio podcast]. American Society for Microbiology. https://asm.org/Podcasts/MTM/Episodes/Microbes,-Heme,-and-Impossible-Burgers-with-Pat-Br

18. Keerie, M. (2021, March 18). *Record $3.1 billion invested in alt proteins in 2020 signals growing market momentum for sustainable proteins.* The Good Food Institute. https://gfi.org/blog/2020-state-of-the-industry-highlights/

19. Ives, M. (2020, December 2). Singapore approves a lab-grown meat product, a global first. *The New York Times.* https://www.nytimes.com/2020/12/02/business/singapore-lab-meat.html

20. Martin, M. J., Thottathil, S. E. & Newman, T. B. (2015). Antibiotics overuse in animal agriculture: A call to action for health care providers. *American Journal of Public Health, 105*(12), 2409–2410. https://doi.org/10.2105/AJPH.2015.302870

21. Theurer, B. M. & Hernandez, A. (2019, August 19). *Carving up the alternative meat market.* Barclays Corporate and Investment Bank. https://www.investmentbank.barclays.com/our-insights/carving-up-the-alternative-meat-market.html

22. World Food Program USA. (2021, October 1). *8 facts to know about food waste and hunger*. https://www.wfpusa.org/articles/8-facts-to-know-about-food-waste-and-hunger/

23. Makkar, H. P. S., Tran, G., Heuzé, V. & Ankers, P. (2014). State-of-the-art on use of insects as animal feed. *Animal Feed Science and Technology, 197*, 1–33. https://doi.org/10.1016/j.anifeedsci.2014.07.008

Chapter 9

1. Loonela, V. & Stoycheva, D. (2021, August 26). *Fishing opportunities in the Baltic for 2022: Improving long-term sustainability of stocks*. European Commission. https://ec.europa.eu/commission/presscorner/detail/en/IP_21_4202

2. Vanninen, P., Östin, A., Bełdowski, J., Pedersen, E. A., Söderström, M., Szubska, M., Grabowski, M., Siedlewicz, G., Czub, M., Popiel, S., Nawała, J., Dziedzic, D., Jakacki, J. & Pączek, B. (2020). Exposure status of sea-dumped chemical warfare agents in the Baltic Sea. *Marine Environmental Research, 161*, 105112. https://doi.org/10.1016/j.marenvres.2020.105112

3. Curry, A. (2016, November 11). Chemical weapons dumped in the ocean after World War II could threaten waters worldwide. *Smithsonian Magazine.* https://www.smithsonianmag.com/science-nature/decaying-weapons-world-war-II-threaten-waters-worldwide-180961046/

4. Chen, S.-C., Sun, G.-X., Yan, Y., Konstantinidis, K. T., Zhang, S.-Y., Deng, Y., Li, X.-M., Cui, H.-L., Musat, F., Popp, D., Rosen, B. P. & Zhu, Y.-G. (2020). The Great Oxidation Event expanded the genetic repertoire of arsenic metabolism and cycling. *Proceedings of the National Academy of Sciences of the United States of America, 117*(19), 10414–10421. https://doi.org/10.1073/pnas.2001063117

5. National Oceanic and Atmospheric Administration. (2021, February 26). *How much oxygen comes from the ocean?* National Ocean Service, U.S.

Department of Commerce. https://oceanservice.noaa.gov/facts/ocean-oxygen.html

6. Chisholm, S. W., Olson, R. J., Zettler, E. R., Goericke, R., Waterbury, J. B. & Welschmeyer, N. A. (1988). A novel free-living prochlorophyte abundant in the oceanic euphotic zone. *Nature, 334*, 340–343. https://doi.org/10.1038/334340a0

7. Breitburg, D., Levin, L. A., Oschlies, A., Grégoire, M., Chavez, F. P., Conley, D. J., Garçon, V., Gilbert, D., Gutiérrez, D., Isensee, K., Jacinto, G. S., Limburg, K. E., Montes, I., Naqvi, S., Pitcher, G. C., Rabalais, N. N., Roman, M. R., Rose, K. A., Seibel, B. A., . . . Zhang, J. (2018). Declining oxygen in the global ocean and coastal waters. *Science, 359*(6371), em7240. https://doi.org/10.1126/science.aam7240

8. Caesar, L., McCarthy, G. D., Thornalley, D. J. R., Cahill, N. & Rahmstorf, S. (2021). Current Atlantic Meridional Overturning Circulation weakest in the last millennium. *Nature Geoscience, 14*, 118–120. https://doi.org/10.1038/s41561-021-00699-z

9. Hellweger, F. L., van Sebille, E., Calfee, B. C., Chandler, J. W., Zinser, E. R., Swan, B. K. & Fredrick, N. D. (2016). The role of ocean currents in the temperature selection of plankton: Insights from an individual-based model. *PLOS ONE, 11*(12), e0167010. https://doi.org/10.1371/journal.pone.0167010

10. United Nations Environment Programme & World Conservation Monitoring Centre. (n.d.). *Climate change and marine diseases: The socio-economic impact.* https://www.unep-wcmc.org/system/dataset_file_fields/files/000/000/125/original/Climate_Change_Marine_Diseases.pdf?1398683242

11. Vezzulli, L., Grande, C., Reid, P. C., Hélaouët, P., Edwards, M., Höfle, M. G., Brettar, I., Colwell, R. R. & Pruzzo, C. (2016). Climate influence on *Vibrio* and associated human diseases during the past half-century in the coastal North Atlantic. *Proceedings of the National Academy of Sciences of the United States of America, 113*(34), E5062–E5071. https://doi.org/10.1073/pnas.1609157113

12. Centers for Disease Control and Prevention. (n.d.). *Questions and answers: Vibrio species causing vibriosis*. https://www.cdc.gov/vibrio/faq.html

13. Deeb, R., Tufford, D., Scott, G. I., Moore, J. G. & Dow, K. (2018). Impact of climate change on *Vibrio vulnificus* abundance and exposure risk. *Estuaries and Coasts: Journal of the Estuarine Research Federation*, *41*(8), 2289–2303. https://doi.org/10.1007/s12237-018-0424-5

14. International Union for Conservation of Nature (IUCN). (2019, December). *Issues Brief: Ocean deoxygenation*. https://www.iucn.org/resources/issues-briefs/ocean-deoxygenation

15. United Nations Educational, Scientific and Cultural Organization (UNESCO). (n.d.). *Facts and figures on marine biodiversity*. http://www.unesco.org/new/en/natural-sciences/ioc-oceans/focus-areas/rio-20-ocean/blueprint-for-the-future-we-want/marine-biodiversity/facts-and-figures-on-marine-biodiversity/

16. Census of Marine Life. (n.d.). *Making ocean life count*. http://www.coml.org/international-census-marine-microbes-icomm/

17. de la Vega, E., Chalk, T. B., Wilson, P. A., Bysani, R. P. & Foster, G. L. (2020). Atmospheric CO_2 during the Mid-Piacenzian Warm Period and the M2 glaciation. *Scientific Reports, 10,* 11002. https://doi.org/10.1038/s41598-020-67154-8

18. The Ocean Portal Team. (2018, April). *Ocean acidification*. Smithsonian Institution. https://ocean.si.edu/ocean-life/invertebrates/ocean-acidification

19. Multi-Agency Rocky Intertidal Network. (2022, February 22). *Sea star wasting syndrome | MARINe*. University of California, Santa Cruz. https://marine.ucsc.edu/data-products/sea-star-wasting/

20. Boyce, D., Lewis, M. & Worm, B. (2010). Global phytoplankton decline over the past century. *Nature, 466*, 591–596. https://doi.org/10.1038/nature09268

21. Ruppel, C. D. & Kessler, J. D. (2017). The interaction of climate change and methane hydrates. *Reviews of Geophysics, 55,* 126–168, https://doi.org/10.1002/2016RG000534

22. Thurber, A. R., Seabrook, S. & Welsh, R. M. (2020). Riddles in the cold: Antarctic endemism and microbial succession impact methane cycling in the Southern Ocean. *Proceedings of the Royal Society B: Biological Sciences, 287*(1931), 20201134. https://doi.org/10.1098/rspb.2020.1134

23. Walworth, N. G., Zakem, E. J., Dunne, J. P., Collins, S. & Levine, N. M. (2020). Microbial evolutionary strategies in a dynamic ocean. *Proceedings of the National Academy of Sciences, 117*(11), 5943–5948. https://www.pnas.org/content/117/11/5943

24. Brown, M. V., van de Kamp, J., Ostrowski, M., Seymour, J. R., Ingleton, T., Messer, L. F., Jeffries, T., Siboni, N., Laverock, B., Bibiloni-Isaksson, J., Nelson, T. M., Coman, F., Davies, C. H., Frampton, D., Rayner, M., Goossen, K., Robert, S., Holmes, B., Abell, G. C. J., . . . Bodrossy, L. (2018). Systematic, continental scale temporal monitoring of marine pelagic microbiota by the Australian Marine Microbial Biodiversity Initiative. *Scientific Data, 5,* 180130. https://doi.org/10.1038/sdata.2018.130

25. McKinsey & Company. (2015, September 1). *Stemming the tide: Land-based strategies for a plastic-free ocean*. https://www.mckinsey.com/business-functions/sustainability/our-insights/stemming-the-tide-land-based-strategies-for-a-plastic-free-ocean

26. Protected Planet. (n.d.). *Marine protected areas*. https://www.protectedplanet.net/en/thematic-areas/marine-protected-areas

27. Ramseur, J. L. (2010, December 16). *Deepwater Horizon oil spill: The fate of the oil*. Congressional Research Service. https://sgp.fas.org/crs/misc/R41531.pdf

28. Yoshida, S., Hiraga, K., Takehana, T., Taniguchi, I., Yamaji, H., Maeda, Y., Toyohara, K., Miyamoto, K., Kimura, Y. & Oda, K. (2016). A bacterium that degrades and assimilates poly(ethylene terephthalate). *Science, 351*(6278), 1196–1199. https://doi.org/10.1126/science.aad6359

29. Anjum, K., Abbas, S. Q., Shah, S. A. A., Akhter, N., Batool, S. & ul Hassan, S. S. (2016). Marine sponges as a drug treasure. *Biomolecules & Therapeutics, 24*(4), 347–362. https://doi.org/10.4062/biomolther.2016.067

30. Zhang, F., Zhao, M., Braun, D. R., Ericksen, S. S., Piotrowski, J. S., Nelson, J., Peng, J., Ananiev, G. E., Chanana, S., Barns, K., Fossen, J., Sanchez, H., Chevrette, M. G., Guzei, I. A., Zhao, C., Guo, L., Tang, W., Currie, C. R., Rajski, S. R., . . . Bugni, T. S. (2020). A marine microbiome antifungal targets urgent-threat drug-resistant fungi. *Science, 370*(6519), 974–978. https://doi.org/10.1126/science.abd6919

Chapter 10

1. Davidson, H. (2021, March 15). Beijing skies turn orange as sandstorm and pollution send readings off the scale. *The Guardian*. https://www.theguardian.com/environment/2021/mar/15/beijing-skies-turn-orange-as-sandstorm-and-pollution-send-readings-off-the-scale
2. Cao, M. & Chen, W. (2019). Epidemiology of lung cancer in China. *Thoracic Cancer, 10*(1), 3–7. https://doi.org/10.1111/1759-7714.12916
3. United Nations. (2019, May 31). *Stressing air pollution kills 7 million people annually, Secretary-General urges governments to build green economy, in message for World Environment Day*. https://www.un.org/press/en/2019/sgsm19607.doc.htm
4. Li, J., Cao, J., Zhu, Y. G., Chen, Q. L., Shen, F., Wu, Y., Xu, S., Fan, H., Da, G., Huang, R. J., Wang, J., de Jesus, A. L., Morawska, L., Chan, C. K., Peccia, J. & Yao, M. (2018). Global survey of antibiotic resistance genes in air. *Environmental Science & Technology, 52*(19), 10975–10984. https://doi.org/10.1021/acs.est.8b02204
5. Tang, Q., Song, P., Li, J., Kong, F., Sun, L. & Xu, L. (2016). Control of antibiotic resistance in China must not be delayed: The current state of resistance and policy suggestions for the government, medical facilities, and patients. *Bioscience Trends, 10*(1), 1–6. https://doi.org/10.5582/bst.2016.01034
6. Chen, J., Wang, Y., Chen, X. & Hesketh, T. (2020). Widespread illegal sales of antibiotics in Chinese pharmacies – A nationwide cross-sectional

study. *Antimicrobial Resistance & Infection Control, 9*(12). https://doi.org/10.1186/s13756-019-0655-7

7. Wang, H., Wang, B., Zhao, Q., Zhao, Y., Fu, C., Feng, X., Wang, N., Su, M., Tang, C., Jiang, F., Zhou, Y., Chen, Y. & Jiang, Q. (2015). Antibiotic body burden of Chinese school children: A multisite biomonitoring-based study. *Environmental Science & Technology, 49*(8), 5070–5079. https://doi.org/10.1021/es5059428
8. World Health Organization. (n.d.). *Antimicrobial resistance in China*. https://www.who.int/china/health-topics/antimicrobial-resistance
9. Mao, Y., Ding, P., Wang, Y., Ding, C., Wu, L., Zheng, P., Zhang, X., Li, X., Wang, L. & Sun, Z. (2019). Comparison of culturable antibiotic-resistant bacteria in polluted and non-polluted air in Beijing, China. *Environment International, 131*, 104936. https://doi.org/10.1016/j.envint.2019.104936
10. Meier, F. C. & Lindbergh, C. A. (1935). Collecting micro-organisms from the arctic atmosphere: With field notes and material. *The Scientific Monthly, 40*(1), 5–20. http://www.jstor.org/stable/15761
11. Kellogg, C. A. & Griffin, D. W. (2006). Aerobiology and the global transport of desert dust. *Trends in Ecology & Evolution, 21*(11), 638–644. https://doi.org/10.1016/j.tree.2006.07.004
12. Lopatina, A., Medvedeva, S., Artamonova, D., Kolesnik, M., Sitnik, V., Ispolatov, Y. & Severinov, K. (2019). Natural diversity of CRISPR spacers of *Thermus*: Evidence of local spacer acquisition and global spacer exchange. *Philosophical Transactions of the Royal Society B: Biological Sciences, 374*(1772), 20180092. https://doi.org/10.1098/rstb.2018.0092
13. Smith, D. J., Ravichandar, J. D., Jain, S., Griffin, D. W., Yu, H., Tan, Q., Thissen, J., Lusby, T., Nicoll, P., Shedler, S., Martinez, P., Osorio, A., Lechniak, J., Choi, S., Sabino, K., Iverson, K., Chan, L., Jaing, C. & McGrath, J. (2018). Airborne bacteria in Earth's lower stratosphere resemble taxa detected in the troposphere: Results from a new NASA Aircraft Bioaerosol Collector (ABC). *Frontiers in Microbiology, 9*, 1752. https://doi.org/10.3389/fmicb.2018.01752

14. Chow, N. A., Griffin, D. W., Barker, B. M., Loparev, V. N. & Litvintseva, A. P. (2016). Molecular detection of airborne *Coccidioides* in Tucson, Arizona. *Medical Mycology, 54*(6), 584–592. https://doi.org/10.1093/mmy/myw022

15. Morrison, D., Crawford, I., Marsden, N., Flynn, M., Read, K., Neves, L., Foot, V., Kaye, P., Stanley, W., Coe, H., Topping, D. & Gallagher, M. (2020). Quantifying bioaerosol concentrations in dust clouds through online UV-LIF and mass spectrometry measurements at the Cape Verde Atmospheric Observatory. *Atmospheric Chemistry and Physics, 20*(22), 14473–14490. https://doi.org/10.5194/acp-20-14473-2020

16. Morawska, L. & Milton, D. K. (2020). It is time to address airborne transmission of coronavirus disease 2019 (COVID-19). *Clinical Infectious Diseases, 71*(9), 2311–2313. https://doi.org/10.1093/cid/ciaa939

17. Fears, A. C., Klimstra, W. B., Duprex, P., Hartman, A., Weaver, S. C., Plante, K. S., Mirchandani, D., Plante, J. A., Aguilar, P. V., Fernández, D., Nalca, A., Totura, A., Dyer, D., Kearney, B., Lackemeyer, M., Bohannon, J. K., Johnson, R., Garry, R. F., Reed, D. S. & Roy, C. J. (2020). Persistence of severe acute respiratory syndrome coronavirus 2 in aerosol suspensions. *Emerging Infectious Diseases, 26*(9), 2168–2171. https://doi.org/10.3201/eid2609.201806

18. Teyssou, E., Delagrèverie, H., Visseaux, B., Lambert-Niclot, S., Brichler, S., Ferre, V., Marot, S., Jary, A., Todesco, E., Schnuriger, A., Ghidaoui, E., Abdi, B., Akhavan, S., Houhou-Fidouh, N., Charpentier, C., Morand-Joubert, L., Boutolleau, D., Descamps, D., Calvez, V., . . . Soulie, C. (2021). The Delta SARS-CoV-2 variant has a higher viral load than the Beta and the historical variants in nasopharyngeal samples from newly diagnosed COVID-19 patients. *The Journal of Infection, 83*(4), E1–E3. https://doi.org/10.1016/j.jinf.2021.08.027

19. World Health Organization. (2021, October 14). *Tuberculosis*. https://www.who.int/news-room/fact-sheets/detail/tuberculosis

20. Magnussen, A. & Parsi, M. A. (2013). Aflatoxins, hepatocellular carcinoma and public health. *World Journal of Gastroenterology, 19*(10), 1508–1512. https://doi.org/10.3748/wjg.v19.i10.1508

21. Ademe, M. & Girma, F. (2020). *Candida auris*: From multidrug resistance to pan-resistant strains. *Infection and Drug Resistance, 13*(2020), 1287–1294. https://doi.org/10.2147/IDR.S249864

22. Federal Bureau of Investigation. (n.d.). *Amerithrax or anthrax investigation*. https://www.fbi.gov/history/famous-cases/amerithrax-or-anthrax-investigation

23. Institute of Medicine (US) and National Research Council (US) Committee on Effectiveness of National Biosurveillance Systems: BioWatch and the Public Health System. (2011). *BioWatch and public health surveillance: Evaluating systems for the early detection of biological threats: Abbreviated version.* National Academies Press. https://doi.org/10.17226/12688

24. U.S. Government Accountability Office. (2021, May 21). *Biodefense: DHS exploring new methods to replace BioWatch and could benefit from additional guidance*. https://www.gao.gov/products/gao-21-292

25. Thomas, N. & Nigam, S. (2018). Twentieth-century climate change over Africa: Seasonal hydroclimate trends and Sahara Desert expansion. *Journal of Climate, 31*(9), 3349–3370. https://doi.org/10.1175/JCLI-D-17-0187.1

Chapter 11

1. Ciavarella, A., Cotterill, D., Stott, P., Kew, S., Philip, S., van Oldenborgh, G. J., Skålevåg, A., Lorenz, P., Robin, Y., Otto, F., Hauser, M., Seneviratne, S. I., Lehner, F. & Zolina, O. (2021). Prolonged Siberian heat of 2020 almost impossible without human influence. *Climatic Change, 166*(9). https://doi.org/10.1007/s10584-021-03052-w

2. Witze, A. (2020). The Arctic is burning like never before — and that's bad news for climate change. *Nature, 585*(7825), 336–337. https://doi.org/10.1038/d41586-020-02568-y

3. Gray, R. (2020, December 1). The mystery of Siberia's exploding craters. *BBC Future*. https://www.bbc.com/future/article/20201130-climate-change-the-mystery-of-siberias-explosive-craters

4. Stella, E., Mari, L., Gabrieli, J., Barbante, C. & Bertuzzo, E. (2020). Permafrost dynamics and the risk of anthrax transmission: A modelling study. *Science Reports, 10*, 16460. https://doi.org/10.1038/s41598-020-72440-6

5. Natali, S. M., Holdren, J. P., Rogers, B. M., Treharne, R., Duffy, P. B., Pomerance, R., & MacDonald, E. (2021). Permafrost carbon feedbacks threaten global climate goals. *Proceedings of the National Academy of Sciences of the United States of America, 118*(21), e2100163118. https://doi.org/10.1073/pnas.2100163118

6. Anthony, K. W., von Deimling, T. S., Nitze, I., Frolking, S., Emond, A., Daanen, R., Anthony, P., Lindgren, P., Jones, B. & Grosse, G. (2018). 21st-century modeled permafrost carbon emissions accelerated by abrupt thaw beneath lakes. *Nature Communications, 9*, 3262. https://doi.org/10.1038/s41467-018-05738-9

7. National Ice and Snow Data Center. (n.d.). *Methane and frozen ground.* https://nsidc.org/cryosphere/frozenground/methane.html

8. Natali, S. M., Watts, J. D., Rogers, B. M., Potter, S., Ludwig, S. M., Selbmann, A.-K., Sullivan, P. F., Abbott, B. W., Arndt, K. A., Birch, L, Björkman, M. P., Bloom, A. A., Celis, G., Christensen, T. R., Christiansen, C. T., Commane, R., Cooper, E. J., Crill, P., Czimczik, C., . . . Zona, D. (2019). Large loss of CO_2 in winter observed across the northern permafrost region. *Nature Climate Change, 9*, 852–857. https://doi.org/10.1038/s41558-019-0592-8

9. McSweeney, R. & Hausfather, Z. (2018, January 15). *Q&A: How do climate models work?* Carbon Brief. https://www.carbonbrief.org/qa-how-do-climate-models-work

10. United Nations Framework Convention on Climate Change. (n.d.). *The Paris Agreement.* https://unfccc.int/process-and-meetings/the-paris-agreement/the-paris-agreement

11. Colatriano, D., Tran, P., Guéguen, C., Williams, W. J., Lovejoy, C. & Walsh, D. A. (2018). Genomic evidence for the degradation of terrestrial organic

matter by pelagic Arctic Ocean Chloroflexi bacteria. *Communications Biology, 1*(90). https://doi.org/10.1038/s42003-018-0086-7

12. Schleussner, C.-F., Lissner, T. K., Fischer, E. M., Wohland, J., Perrette, M., Golly, A., Rogelj, J., Childers, K., Schewe, J., Frieler, K., Mengel, M., Hare, W. & Schaeffer, M. (2016). Differential climate impacts for policy-relevant limits to global warming: The case of 1.5 °C and 2 °C. *Earth System Dynamics, 7* (2), 327–351. https://doi.org/10.5194/esd-7-327-2016

13. IPCC. (2021, August 7). *Climate change 2021: The physical science basis. Contribution of Working Group I to the Sixth Assessment Report of the Intergovernmental Panel on Climate Change* [V. Masson-Delmotte, P. Zhai, A. Pirani, S. L. Connors, C. Péan, S. Berger, N. Caud, Y. Chen, L. Goldfarb, M. I. Gomis, M. Huang, K. Leitzell, E. Lonnoy, J. B. R. Matthews, T. K. Maycock, T. Waterfield, O. Yelekçi, R. Yu & B. Zhou (Eds.)]. Cambridge University Press. https://www.ipcc.ch/report/ar6/wg1/

14. United States Environmental Protection Agency. (2021, July 21). *Climate change indicators: Atmospheric concentrations of greenhouse gases*. https://www.epa.gov/climate-indicators/climate-change-indicators-atmospheric-concentrations-greenhouse-gases

15. Brienen, R. J. W., Phillips, O. L., Feldpausch, T. R., Gloor, E., Baker, T. R., Lloyd, J., Lopez-Gonzalez, G., Monteagudo-Mendoza, A., Malhi, Y., Lewis, S. L., Vásquez Martinez, R., Alexiades, M., Álvarez Dávila, E., Alvarez-Loayza, P., Andrade, A., Aragão, L. E. O. C., Araujo-Murakami, A., Arets, E. J. M. M., Arroyo, L., . . . Zagt, R. J. (2015). Long-term decline of the Amazon carbon sink. *Nature, 519*, 344–348. https://doi.org/10.1038/nature14283

16. Covey, K., Soper, F., Pangala, S., Bernardino, A., Pagliaro, Z., Basso, L., Cassol, H., Fearnside, P., Navarrete, D., Novoa, S., Sawakuchi, H., Lovejoy, T., Marengo, J., Peres, C., Baillie, J., Bernasconi, P., Camargo, J., Freitas, C., Hoffman, B., . . . Elmore, A. (2021). Carbon and beyond: The biogeochemistry of climate in a rapidly changing Amazon. *Frontiers in Forests and Global Change, 4*, 618401. https://doi.org/10.3389/ffgc.2021.618401

17. Kroeger, M. E., Meredith, L. K., Meyer, K. M., Webster, K. D., de Camargo, P. B., de Souza, L. F., Tsai, S. M., van Haren, J., Saleska, S., Bohannan,

B. J. M., Rodrigues, J. L. M., Berenguer, E., Barlow, J. & Nüsslein, K. (2021). Rainforest-to-pasture conversion stimulates soil methanogenesis across the Brazilian Amazon. *The ISME Journal, 15*, 658–672. https://doi.org/10.1038/s41396-020-00804-x

18. Puxty, R. J., Millard, A. D., Evans, D. J & Scanlan, D. J. (2016). Viruses inhibit CO_2 fixation in the most abundant phototrophs on Earth. *Current Biology, 26*(12), 1585–1589. https://doi.org/10.1016/j.cub.2016.04.036
19. RISING Global Peace Forum. (2020, November 14). *Lord Mayor's Peace Lecture 2020 — Sir David King* [Video]. YouTube. https://www.youtube.com/watch?v=IB4e1Y9C9oc

Chapter 12

1. Stokes, J. M., Yang, K., Swanson, K., Jin, W., Cubillos-Ruiz, A., Donghia, N. M., MacNair, C. R., French, S., Carfrae, L. A., Bloom-Ackermann, Z., Tran, V. M., Chiappino-Pepe, A., Badran, A. H., Andrews, I. W., Chory, E. J., Church, G. M., Brown, E. D., Jaakkola, T. S., Barzilay, R. & Collins, J. J. (2020). A deep learning approach to antibiotic discovery. *Cell, 180*(4), 688–702. https://doi.org/10.1016/j.cell.2020.01.021
2. The National Security Commission on Artificial Intelligence. (2021, March). *Final report: National Security Commission on Artificial Intelligence*. https://www.nscai.gov/wp-content/uploads/2021/03/Full-Report-Digital-1.pdf
3. Cameron, D. E., Bashor, C. J. & Collins, J. J. (2014). A brief history of synthetic biology. *Nature Reviews Microbiology, 12*(5), 381–390. https://doi.org/10.1038/nrmicro3239
4. Moore, G. E. (1965). *The future of integrated electronics*. https://archive.computerhistory.org/resources/access/text/2017/03/102770836-05-01-acc.pdf
5. Ro, D.-K., Paradise, E. M., Ouellet, M., Fisher, K. J., Newman, K. L., Ndungu, J. M., Ho, K. A., Eachus, R. A., Ham, T. S., Kirby, J., Chang,

M. C. Y., Withers, S. T., Shiba, Y., Sarpong, R. & Keasling, J. D. (2006). Production of the antimalarial drug precursor artemisinic acid in engineered yeast. *Nature, 440*, 940–943. https://doi.org/10.1038/nature04640

6. BioBricks Foundation. (n.d.). *Building with biology to benefit all people and the planet*. https://biobricks.org
7. Dunn, S.-J. (2019, July). *The next software revolution: Programming biological cells* [Video]. Ted Talk. https://www.ted.com/talks/sara_jane_dunn_the_next_software_revolution_programming_biological_cells?language=en
8. Chui, M., Evers, M., Manyika, J., Zheng, A. & Nisbet, T. (2020, May 13). *The Bio Revolution: Innovations transforming economies, societies, and our lives*. McKinsey & Company. https://www.mckinsey.com/industries/life-sciences/our-insights/the-bio-revolution-innovations-transforming-economies-societies-and-our-lives
9. Collins, N. (2019, August 27). *Stanford researchers probe the link between gut microbiota and cancer immunotherapy*. Standford University. https://news.stanford.edu/2019/08/27/researchers-probe-microbiome-cancer-treatment-link/
10. Moradali, M. F. & Rehm, B. H. A. (2020). Bacterial biopolymers: From pathogenesis to advanced materials. *Nature Reviews Microbiology, 18*, 195–210. https://doi.org/10.1038/s41579-019-0313-3
11. Goldman, N., Bertone, P., Chen, S., Dessimoz, C., LeProust, E. M., Sipos, B. & Birney, E. (2013). Towards practical, high-capacity, low-maintenance information storage in synthesized DNA. *Nature, 494*(7435), 77–80. https://doi.org/10.1038/nature11875
12. Rydning, J. & Shirer, M. (2020, May 8). *IDC's Global DataSphere forecast shows continued steady growth in the creation and consumption of data*. Business Wire. https://www.businesswire.com/news/home/20200508005025/en/
13. Quirks & Quarks. (2017, November 10). *Meet the human guinea pig who hacked his own DNA* [Radio broadcast]. CBC Radio. https://www.cbc.ca/radio/quirks/diy-dna-hacks-wounds-take-longer-to-heal-at-night-why-daydreams-are-good-quirks-bombs-and-more-1.4395576/meet-the-human-

guinea-pig-who-hacked-his-own-dna-1.4395589?fbclid=IwAR3zMp6Cb c9cU7_YNVb7jwuQz3wsaqZhWATWlIgg7IrBt8zqTbTCD6pGFmg

14. Balakrishnan, V. S. (2018, October 2). Celebrated Brazilian bee scientist Warwick Kerr dies: Revered as a humanitarian and scientist, Kerr was also blamed for the introduction of aggressive Africanized bees to the Americas. *The Scientist.* https://www.the-scientist.com/news-opinion/celebrated-brazilian-bee-scientist-warwick-kerr-dies-64886

15. Centers for Disease Control and Prevention. (2019). *QuickStats*: Number of deaths from hornet, wasp, and bee stings, among males and females — National Vital Statistics System, United States, 2000–2017. *Morbidity & Mortality Weekly Report, 68*(29), 649. https://www.cdc.gov/mmwr/volumes/68/wr/mm6829a5.htm#:~:text=During%202000%E2%80%932017%2C%20a%20total,the%20deaths%20were%20among%20males

16. Cyranoski, D. (2019). The CRISPR-baby scandal: What's next for human gene-editing. *Nature, 566*, 440–442. https://doi.org/10.1038/d41586-019-00673-1

17. Cyranoski D. (2019). Russian biologist plans more CRISPR-edited babies. *Nature, 570*, 145–146. https://doi.org/10.1038/d41586-019-01770-x

18. Ishiguro, K. (2005). *Never let me go*. Faber and Faber Limited.

19. Klymiuk, N., Aigner, B., Brem, G., & Wolf, E. (2010). Genetic modification of pigs as organ donors for xenotransplantation. *Molecular Reproduction and Development*, *77*(3), 209–221. https://doi.org/10.1002/mrd.21127

20. Ratcliffe, J. (2020, December 3). China is national security threat no. 1. *The Wall Street Journal.* https://www.wsj.com/articles/china-is-national-security-threat-no-1-11607019599

21. Emanuel, P., Walper, S., DiEuliis, D., Klein, N., Petro, J. B. & Giordano, J. (2019, October). *Cyborg soldier 2050: Human/Machine fusion and the implications for the future of the DoD*. U.S. Army Combat Capabilities Development Command Chemical Biological Center. https://apps.dtic.mil/sti/citations/AD1083010

22. Cunningham, M. A. & Geis, J. P., II. (2020). A national strategy for synthetic biology. *Strategic Studies Quarterly, 14*(3), 49–80. https://www.jstor.org/stable/26937411

Chapter 13

1. McLeish, C. (2010). Opening up the secret city of Stepnogorsk: Biological weapons in the Former Soviet Union. *Area, 42*(1), 60–69. http://www.jstor.org/stable/27801440
2. Weber, A. (2017). *Countering weapons of mass destruction without a map* [Video]. Ted Talk. https://www.ted.com/talks/andrew_weber_countering_weapons_of_mass_destruction_without_a_map
3. United Nations. (n.d.). *Biological Weapons Convention*. https://www.un.org/disarmament/biological-weapons/
4. United Nations. (n.d.). *UN Security Council Resolution 1540 (2004).* https://www.un.org/disarmament/wmd/sc1540/
5. National Academies of Sciences, Engineering, and Medicine. (2018). *Biodefense in the age of synthetic biology.* National Academies Press. https://doi.org/10.17226/24890
6. Centers for Disease Control and Prevention. (n.d.). *Select agents and toxins list*. https://www.selectagents.gov/sat/list.htm
7. Federal Bureau of Investigation. (n.d.). *Amerithrax or anthrax investigation*. https://www.fbi.gov/history/famous-cases/amerithrax-or-anthrax-investigation
8. Wheelis, M. (2002). Biological warfare at the 1346 siege of Caffa. *Emerging Infectious Diseases, 8*(9), 971–975. https://doi.org/10.3201/eid0809.010536
9. Christopher, G. W., Cieslak, T. J., Pavlin, J. A. & Eitzen, E. M., Jr. (1997). Biological warfare. A historical perspective. *JAMA, 278*(5), 412–417. https://pubmed.ncbi.nlm.nih.gov/9244333/

10. Bradsher, B. (n.d.). *Japanese war crimes and related records: A guide to records in the national archives*. The U.S. National Archives and Records Administration. https://www.archives.gov/iwg/japanese-war-crimes

11. U.S. Government Accountability Office. (2004, May 14). *Chemical and biological defense: DOD needs to continue to collect and provide information on tests and on potentially exposed personnel*. https://www.gao.gov/products/gao-04-410

12. Noyce, R. S., Lederman, S. & Evans, D. H. (2018). Construction of an infectious horsepox virus vaccine from chemically synthesized DNA fragments. *PLOS ONE, 13*(1), e0188453. https://doi.org/10.1371/journal.pone.0188453

13. Callaway E. (2020). 'It will change everything': DeepMind's AI makes gigantic leap in solving protein structures. *Nature*, *588*, 203–204. https://doi.org/10.1038/d41586-020-03348-4

14. Marks, M. (2017, December 19). *Inside Texas' anthrax triangle* [Audio podcast]. Texas Standard. https://www.texasstandard.org/stories/inside-texas-anthrax-triangle/

15. Aldrich, J. L. (2003). One laboratory scientist's experience as part of the UN team inspecting iraq's biowarfare capability. *Bios*, *74*(1), 22–24. http://www.jstor.org/stable/4608662

16. Clapper, J. R. (2016, February 9). *Worldwide threat assessment of the US Intelligence Community*. https://www.dni.gov/files/documents/SASC_Unclassified_2016_ATA_SFR_FINAL.pdf

17. Zhen, L. (2019, December 14). Chinese criminal gangs spreading African swine fever to force farmers to sell pigs cheaply so they can profit. *South China Morning Post*. https://www.scmp.com/news/china/politics/article/3042122/chinese-criminal-gangs-spreading-african-swine-fever-force

18. Broniatowski, D. A., Jamison, A. M., Qi, S., AlKulaib, L., Chen, T., Benton, A., Quinn, S. C. & Dredze, M. (2018). Weaponized health communication: Twitter bots and Russian trolls amplify the vaccine debate. *American Journal of Public Health, 108*(10), 1378–1384. https://doi.org/10.2105/AJPH.2018.304567

19. Kitfield, J. (2021, March 11). ‘We’re going to lose fast’: U.S. Air Force held a war game that started with a Chinese biological attack. *Yahoo! News*. https://news.yahoo.com/were-going-to-lose-fast-us-air-force-held-a-war-game-that-started-with-a-chinese-biological-attack-170003936.html?__twitter_impression=true

20. U.S. Department of State. (2021, January 15). *Fact sheet: Activity at the Wuhan Institute of Virology*. https://2017-2021.state.gov/fact-sheet-activity-at-the-wuhan-institute-of-virology/index.html

21. Qiu, J. (2020, June 1). How China’s ‘bat woman’ hunted down viruses from SARS to the new coronavirus. *Scientific American*. https://www.scientificamerican.com/article/how-chinas-bat-woman-hunted-down-viruses-from-sars-to-the-new-coronavirus1/

22. Selgelid, M. J. (2016). Gain-of-function research: Ethical analysis. *Science and Engineering Ethics, 22*(4), 923–964. https://doi.org/10.1007/s11948-016-9810-1

23. Kaiser, J. (2014, July 11). Lab incidents lead to safety crackdown at CDC: Agency halts shipments from high-containment labs, announces reforms in wake of lapses involving anthrax, smallpox, and influenza. *Science Insider*. https://www.science.org/news/2014/07/lab-incidents-lead-safety-crackdown-cdc

24. Collins, F. S. (2014, October 16). *Statement on funding pause on certain types of gain-of-function research*. National Institutes of Health. https://www.nih.gov/about-nih/who-we-are/nih-director/statements/statement-funding-pause-certain-types-gain-function-research

25. Lentzos, F. & Koblentz, G. (2021, June 14). *Fifty-nine labs around world handle the deadliest pathogens — only a quarter score high on safety*. King’s College London. https://www.kcl.ac.uk/fifty-nine-labs-around-world-handle-the-deadliest-pathogens-only-a-quarter-score-high-on-safety

26. Centers for Disease Control and Prevention. (n.d.). *History of smallpox*. https://www.cdc.gov/smallpox/history/history.html

27. Zhong, N. & Zeng, G. (2006). What we have learnt from SARS epidemics in China. *The BMJ, 333*(7564), 389–391. https://doi.org/10.1136/bmj.333.7564.389

28. Grady, D. (2019, August 5). Deadly germ research is shut down at army lab over safety concerns. *The New York Times.* https://www.nytimes.com/2019/08/05/health/germs-fort-detrick-biohazard.html

Chapter 14

1. Schopf, J. W., Kitajima, K., Spicuzza, M. J., Kudryavtsev, A. B. & Valley, J. W. (2018). SIMS analyses of the oldest known assemblage of microfossils document their taxon-correlated carbon isotope compositions. *Proceedings of the National Academy of Sciences of the United States of America, 115*(1), 53–58. https://doi.org/10.1073/pnas.1718063115

2. Valley, J. W., Cavosie, A. J., Ushikubo, T., Reinhard, D. A., Lawrence, D. F., Larson, D. J., Clifton, P. H., Kelly, T. F., Wilde, S. A., Moser, D. E. & Spicuzza, M. J. (2014). Hadean age for a post-magma-ocean zircon confirmed by atom-probe tomography. *Nature Geoscience, 7*, 219–223. https://doi.org/10.1038/ngeo2075

3. Darwin, C. (1871, February 1). *Darwin correspondence project, "letter no. 7471".* Cambridge University Library. https://www.darwinproject.ac.uk/letter/DCP-LETT-7471.xml

4. Urey, H. C. (1952). On the early chemical history of the earth and the origin of life. *Proceedings of the National Academy of Sciences of the United States of America, 38*(4), 351–363. https://doi.org/10.1073/pnas.38.4.351

5. Watson, J. D. & Crick, F. H. C. (1953). Molecular structure of nucleic acids: A structure for deoxyribose nucleic acid. *Nature, 171*(4356), 737–738. https://doi.org/10.1038/171737a0

6. Weiss, M. C., Preiner, M., Xavier, J. C., Zimorski, V. & Martin, W. F. (2018). The Last Universal Common Ancestor between ancient Earth chemistry and the onset of genetics. *PLOS Genetics, 14*(8), e1007518. https://doi.org/10.1371/journal.pgen.1007518

7. Darwin, C. (1859). *On the origin of species by means of natural selection, or preservation of favoured races in the struggle for life.* John Murray.
8. Goldenfeld, N. (2014). Looking in the right direction: Carl Woese and evolutionary biology. *RNA Biology, 11*(3), 248–253. https://doi.org/10.4161/rna.28640
9. Woese, C. R. & Goldenfeld, N. (2009). How the microbial world saved evolution from the scylla of molecular biology and the charybdis of the modern synthesis. *Microbiology and Molecular Biology Reviews, 73*(1), 14–21. https://doi.org/10.1128/MMBR.00002-09
10. Woese, C. R. & Fox, G. E. (1977). Phylogenetic structure of the prokaryotic domain: The primary kingdoms. *Proceedings of the National Academy of Sciences of the United States of America, 74*(11), 5088–5090. https://doi.org/10.1073/pnas.74.11.5088
11. Mayr, E. (1998). Two empires or three? *Proceedings of the National Academy of Sciences of the United States of America, 95*(17), 9720–9723. https://doi.org/10.1073/pnas.95.17.9720
12. Morell, V. (1997). Microbial biology: Microbiology's scarred revolutionary. *Science, 276*(5313), 699–702. https://doi.org/10.1126/science.276.5313.699
13. Goldenfeld, N. & Woese, C. (2007). Biology's next revolution. *Nature, 445*(369). https://doi.org/10.1038/445369a
14. Sun, D., Jeannot, K., Xiao, Y. & Knapp, C. W. (2019). Editorial: Horizontal gene transfer mediated bacterial antibiotic resistance. *Frontiers in Microbiology, 10*, 1933. https://doi.org/10.3389/fmicb.2019.01933
15. van Gestel, J., Bareia, T., Tenennbaum, B., Dal Co, A., Guler, P., Aframian, N., Puyesky, S., Grinberg, I., D'Souza, G. G., Erez, Z., Ackermann, M. & Eldar, A. (2021). Short-range quorum sensing controls horizontal gene transfer at micron scale in bacterial communities. *Nature Communications, 12*, 2324. https://doi.org/10.1038/s41467-021-22649-4
16. Enard, D., Cai, L., Gwennap, C. & Petrov, D. A. (2016). Viruses are a dominant driver of protein adaptation in mammals. *eLife, 5*, e12469. https://doi.org/10.7554/eLife.12469

17. Nobel Prize Outreach AB 2021. (n.d.). *The Nobel Prize in Chemistry 1989.* NobelPrize.org. https://www.nobelprize.org/prizes/chemistry/1989/summary/

18. MoleCluesTV. (2019, May 16). *Jack Szostak: The early Earth and the origins of cellular life* [Video]. YouTube. https://www.youtube.com/watch?v=h-1KqvoLEj7c

19. England, J. L. (2013). Statistical physics of self-replication. *The Journal of Chemical Physics, 139*(12). https://doi.org/10.1063/1.4818538

20. Schrödinger, E. (1944). *What is life? The physical aspect of the living cell.* Cambridge University Press.

21. Horowitz, J. M. & England, J. L. (2017). Spontaneous fine-tuning to the environment in many-species chemical reaction networks. *Proceedings of the National Academy of Sciences of the United States of America, 114*(29), 7565–7570. https://doi.org/10.1073/pnas.1700617114

22. English, J. (2020, November 1). Reconsidering life's origin. *The Scientist.* https://www.the-scientist.com/reading-frames/reconsidering-lifes-origin-68064

Chapter 15

1. Ozaki, K. & Reinhard, C. T. (2021). The future lifespan of Earth's oxygenated atmosphere. *Nature Geoscience, 14*, 138–142. https://doi.org/10.1038/s41561-021-00693-5

2. Schröder, K.-P. & Smith, R. C. (2008). Distant future of the Sun and Earth revisited. *Monthly Notices of the Royal Astronomical Society, 386*(1), 155–163. https://doi.org/10.1111/j.1365-2966.2008.13022.x

3. Klein, H. P. (1978). The Viking biological experiments on Mars. *Icarus, 34*(3), 666–674. https://doi.org/10.1016/0019-1035(78)90053-2

4. NASA Exoplanet Exploration. (2021). *Latest data from NASA's exoplanet archive.* https://exoplanets.nasa.gov/discovery/exoplanet-catalog/

5. Kawaguchi, Y., Shibuya, M., Kinoshita, I., Yatabe, J., Narumi, I., Shibata, H., Hayashi, R., Fujiwara, D., Murano, Y., Hashimoto, H., Imai, E., Kodaira, S., Uchihori, Y., Nakagawa, K., Mita, H., Yokobori, S.-i. & Yamagishi, A. (2020). DNA damage and survival time course of Deinococcal cell pellets during 3 years of exposure to outer space. *Frontiers in Microbiology, 11*, 2050. https://doi.org/10.3389/fmicb.2020.02050

6. NASA Science. (2021). *Mars 2020 mission: Perseverance rover*. https://mars.nasa.gov/mars2020/

7. United Nations Office for Outer Space Affairs. (2021). *Treaty on principles governing the activities of states in the exploration and use of outer space, including the moon and other celestial bodies*. https://www.unoosa.org/oosa/en/ourwork/spacelaw/treaties/introouterspacetreaty.html

8. McKay, C. P., Stoker, C. R., Glass, B. J., Davé, A. I., Davila, A. F., Heldmann, J. L., Marinova, M. M., Fairen, A. G., Quinn, R. C., Zacny, K. A., Paulsen, G., Smith, P. H., Parro, V., Andersen, D. T., Hecht, M. H., Lacelle, D. & Pollard, W. H. (2013). The Icebreaker Life mission to Mars: A search for biomolecular evidence for life. *Astrobiology*, *13*(14), 334–353. https://doi.org/10.1089/ast.2012.0878

9. Fairén, A. G., Parro, V., Schulze-Makuch, D. & Whyte, L. (2017). Searching for life on mars before it is too late. *Astrobiology*, *17*(10), 962–970. https://doi.org/10.1089/ast.2017.1703

10. Greaves, J. S., Richards, A. M. S., Bains, W., Rimmer, P. B., Sagawa, H., Clements, D. L., Seager, S., Petkowski, J. J., Sousa-Silva, C., Ranjan, S., Drabek-Maunder, E., Fraser, H. J., Cartwright, A., Mueller-Wodarg, I., Zhan, Z., Friberg, P., Coulson, I., Lee, E. & Hoge, J. (2021). Phosphine gas in the cloud decks of Venus. *Nature Astronomy, 5*, 655–664. https://doi.org/10.1038/s41550-020-1174-4

11. Steigerwald, W. & Jones, N. N. (2021, June 3). *NASA to explore divergent fate of Earth's mysterious twin with Goddard's DAVINCI+*. NASA. https://www.nasa.gov/feature/goddard/2021/nasa-to-explore-divergent-fate-of-earth-s-mysterious-twin-with-goddard-s-davinci

12. Savage, D. L., Hartsfield, J. & Salisbury, D. (1996, August 7). *Meteorite yields evidence of primitive life on early Mars*. NASA Jet Propulsion Laboratory. https://www2.jpl.nasa.gov/snc/nasa1.html

13. NASA Jet Propulsion Laboratory. (2021). *Europa Clipper: NASA's Europa Clipper will conduct detailed reconnaissance of Jupiter's moon Europa and investigate whether the icy moon could have conditions suitable for life*. https://www.jpl.nasa.gov/missions/europa-clipper

14. NASA. (2020, September). *The Artemis plan: NASA's lunar exploration program overview*. https://www.nasa.gov/sites/default/files/atoms/files/artemis_plan-20200921.pdf

15. Sagan, C. (1961). The planet Venus: Recent observations shed light on the atmosphere, surface, and possible biology of the nearest planet. *Science, 133* (3456), 849–858. https://doi.org/10.1126/science.133.3456.849

16. Sagan, C. (1971, August). *The long winter model of martian biology: A speculation*. Presented at the Viking Annual Science Seminar, NASA Langley Research Center, 20 April, 1971. https://ntrs.nasa.gov/api/citations/19720005188/downloads/19720005188.pdf

17. The Late Show with Stephen Colbert. (2015, September 10). *Elon Musk might be a super villain* [Video]. YouTube. https://youtu.be/gV6hP-9wpMW8

18. Lopez, J. V., Peixoto, R. S. & Rosado, A. S. (2019). Inevitable future: space colonization beyond Earth with microbes first. *FEMS Microbiology Ecology, 95*(10), fiz127. https://doi.org/10.1093/femsec/fiz127

19. Jakosky, B. M. & Edwards, C. S. (2018). Inventory of CO_2 available for terraforming Mars. *Nature Astronomy, 2*, 634–639. https://doi.org/10.1038/s41550-018-0529-6

20. Wordsworth, R., Kerber, L. & Cockell, C. (2019). Enabling Martian habitability with silica aerogel via the solid-state greenhouse effect. *Nature Astronomy, 3*, 898–903. https://doi.org/10.1038/s41550-019-0813-0

21. Cockell, C. S., Santomartino, R., Finster, K., Waajen, A. C., Nicholson, N., Loudon, C. M., Eades, L. J., Moeller, R., Rettberg, P., Fuchs, F. M., Van Houdt, R., Leys, N., Coninx, I., Hatton, J., Parmitano, L., Krause, J., Koehler, A., Caplin, N., Zuijderduijn, L., Mariani, A., . . . Demets, R. (2021). Microbially-enhanced vanadium mining and bioremediation under micro- and Mars gravity on the International Space Station. *Frontiers in Microbiology, 12*, 641387. https://doi.org/10.3389/fmicb.2021.641387

22. Lubin, P. (2016). A roadmap to interstellar flight. *Journal of the British Interplanetary Society*, *69*(2), 40–72. https://arxiv.org/abs/1604.01356

23. Lubin, P. (2020, July). *How humanity can reach the stars* [Video]. Ted Talk. https://www.ted.com/talks/philip_lubin_how_humanity_can_reach_the_stars/transcript?language=en

24. Perry, L. (2020, May 15). *FLI podcast: On the future of computation, synthetic biology, and life with George Church* [Audio podcast]. Future of Life Institute. https://futureoflife.org/2020/05/15/on-the-future-of-computation-synthetic-biology-and-life-with-george-church/

25. Gros, C. (2016). Developing ecospheres on transiently habitable planets: The Genesis project. *Astrophysics and Space Science, 361*(324). https://doi.org/10.1007/s10509-016-2911-0

Conclusion

1. Kuhn, T. S. (1962). *The structure of scientific revolutions.* The University of Chicago Press.

2. Woese, C. R. & Goldenfeld, N. (2009). How the microbial world saved evolution from the scylla of molecular biology and the charybdis of the modern synthesis. *Microbiology and Molecular Biology Reviews, 73*(1), 14–21. https://doi.org/10.1128/MMBR.00002-09

3. Wilson, E. O. (1992). *The diversity of life.* Belknap Press of Harvard University Press.

4. Craven, M., Sabow, A., Van der Veken, L. & Wilson, M. (2021, May 21). *Not the last pandemic: Investing now to reimagine public-health systems.* McKinsey & Company. https://www.mckinsey.com/industries/public-and-social-sector/our-insights/not-the-last-pandemic-investing-now-to-reimagine-public-health-systems

5. Sirleaf, E. J. & Clark, H. (2021). Report of the Independent Panel for Pandemic Preparedness and Response: Making COVID-19 the last pandemic. *The Lancet, 398*(10295), 101–103. https://doi.org/10.1016/S0140-6736(21)01095-3

6. Baedke, J., Fábregas-Tejeda, A. & Nieves Delgado, A. (2020). The holobiont concept before Margulis. *Journal of Experimental Zoology Part B: Molecular and Developmental Evolution, 334*(3). 149–155. https://doi.org/10.1002/jez.b.22931

7. Gould, S. J. (1996, November 13). Planet of the bacteria. *The Washington Post.* https://www.washingtonpost.com/archive/1996/11/13/planet-of-the-bacteria/6fb60f1d-e6fe-471e-8a0f-4cfa9373772c/

Index